The Crisis Within

America's Military and the Struggle Between the Overseas and Guardian Paradigms

Paula G. Thornhill

Prepared for the United States Air Force
Approved for public release; distribution unlimited

For more information on this publication, visit www.rand.org/t/RR1420

Library of Congress Cataloging-in-Publication Data
is available for this publication.

ISBN: 978-0-8330-9300-4

Published by the RAND Corporation, Santa Monica, Calif.

Preface

At some point during their military service, almost everyone who wears a uniform wonders about the gap between the military they thought they joined and the reality of the military in which they serve. Since 2001, these conversations seem to have occurred with increased frequency, and they suggest an ever-growing dissonance between the theory and reality of military service. This report represents an intellectual journey that several RAND, military, and government civilian colleagues have taken with me to understand and explain this gap. The report is offered as a simple departure point for a critical discussion about the nature and purpose of America's military in the 21st century. If it contributes in any small way to that discussion, it will have achieved its purpose. This work was conducted within RAND Project AIR FORCE.

RAND Project AIR FORCE

RAND Project AIR FORCE (PAF), a division of the RAND Corporation, is the U.S. Air Force's federally funded research and development center for studies and analyses. PAF provides the Air Force with independent analyses of policy alternatives affecting the development, employment, combat readiness, and support of current and future air, space, and cyber forces. Research is conducted in four programs: Force Modernization and Employment; Manpower, Personnel, and Training; Resource Management; and Strategy and Doctrine. The research reported here was prepared under contract FA7014-06-C-0001.

Additional information about PAF is available on our website: www.rand.org/paf/

Contents

Summary

Today's U.S. military is full of perplexing questions and issues. Individually, each can be explained, but collectively the explanations seem too complicated. This complexity makes the military difficult to comprehend, even to those in uniform. What follows is an attempt to unravel this complexity and to start a conversation about how to better understand America's 21st-century military. To do that, a return to first principles is necessary, starting with how the nation understands "the common defense" and the role that organized violence plays in providing for it. The nation's understanding of both the common defense and organized violence changed dramatically in the years since 2001. The diversification in the employment of violence produced a profound paradigm shift that Thomas Kuhn's seminal work, *The Structure of Scientific Revolutions*, helps to identify and explain. America's senior civilian and military leaders must understand this shift in order to create the military the nation needs in the coming decades and to ensure that it is an institution the American people continue to trust.

Acknowledgments

Several mentors, colleagues, and friends were instrumental in bringing this report to fruition. A few people in particular deserve mention: Gen Richard B. Myers (Ret.), Gen Douglas Fraser (Ret.), ADM James Stavridis (Ret.), Maj Gen Jeff Newell, CAPT Scott Smith, Col Jeff Kojac, LTC Buzz Phillips, and Maj Julia Muedeking helped to clarify the argument from a military perspective. Daniel Moran from the Naval Postgraduate School and RAND colleagues Lisa Harrington, Andy Hoehn, David Johnson, Lara Schmidt, and Alan Vick greatly enhanced the quality of this report with thoughtful comments and reviews. My assistant, Lt Col Kari Thyne (Ret.), a keen scholar in her own right, commented on several drafts of the document, greatly improving the argument and insights in the process. Finally, Mara Karlin, the embodiment of the policymaker-scholar nexus, repeatedly asked about the report, encouraging me to stay with the inquiry for the long haul. Any shortcomings in this report reflect my inability to capture the key points all made so patiently and ably.

Prologue: The Colonel's Crisis

The colonel hid it well, but he was facing an existential crisis. He had a speech to give—his retirement speech. After 26 years in the military, he was retiring in a few days. The speech should have been easy to write. The colonel joined the military in the early 1990s, inspired by his father and grandfather. He'd gone through ROTC at an excellent state university and joined the military immediately after graduation. The colonel had always been proud of his service. He felt that he was carrying on the family tradition; his grandfather had served as an enlisted man during the Second World War, and his father likewise in Vietnam. They weren't career men, but they were proud of their time overseas, proud of being in harm's way, proud of their service.

Yet, as the colonel set out to write, he grew perplexed. He wanted to highlight his multigenerational military connection, but how? He couldn't use personal sacrifice or physical hardship. His father and grandfather had spent most of their time in uniform overseas and in war zones, paid the low salaries of America's pre–all-volunteer force. The colonel, on the other hand, had been compensated nicely for his unique skills and had spent his entire career stationed in the United States. Indeed, he spent so much time in one state that he realized he moved less than his college classmates in the civilian world. He also lived in the same home, in the same neighborhood, in the same city for so many years that his neighbors frequently forgot that he was in the military.

He couldn't use the uniqueness of his military organization and the associated camaraderie of brothers-in-arms. His father and grandfather told stories about how their units bonded under dangerous conditions, where the needs of the military unit defined their individual needs. The colonel, however, had spent much of his career on the periphery—even outside the military organization he joined in ROTC—working side by side more with civilians than those in uniform. Initially, he had even felt some disdain for these civilians—convinced that those in uniform did the important, operational work. But, as he approached retirement, his attitude changed. In fact, if he were honest with himself, he had more in common with these civilians than he did with some of his war college classmates, much less his father and grandfather. Indeed, in a few weeks, he'd be back in his old organization, doing the same sort of work he had for decades, but this time wearing khakis and a polo shirt.

And, most distressing of all, he couldn't use proximity to physical danger, the hallmark of military service. The colonel had never even traveled to a war zone, but his younger sister had, and she paid the ultimate price. But she died as a contractor, not a service member. Looking for adventure, his sister had joined a large security contractor supporting operations overseas. She never came home; she was another victim of an IED attack. Yet, beyond the obligatory obituary in the local newspaper, few outside the family remarked about her death. To add to his distress, when one of her high school classmates who joined the military was killed in almost the identical location a few weeks later, the entire town turned out to mourn and to witness the posthumous award of the Purple Heart medal. No, to raise any such comparison about shared danger would diminish what his father and grandfather had done, and even mock his sister's sacrifice. He couldn't do it.

What was going on? The more the colonel tried to make connections back to the military of his father and grandfather, the more artificial and brittle they seemed. He had volunteered, taken an oath, and worn a uniform, but was that enough? After all these years, in the final days of his career, he was distraught. Had he actually been in the military at all?

Introduction

At the individual level, the experiences of this hypothetical colonel highlight some of the large, profound issues affecting America's 21st-century military. As the U.S. military evolved in the years since the nation's founding, the building-block organizations remained the military services: the Army, Navy, Marine Corps, and, since 1947, the Air Force. But, over time, the services' unique purposes increasingly blurred, along with the roles and identities of those serving in the ranks. Originally, the framers of the Constitution conceived of military organizations devoted to the common defense and composed of personnel required to conduct land and sea operations under isolated circumstances, removed from close contact with the government or other citizens. In the intervening centuries, though, this founding requirement morphed into something quite different. For example, some service members, such as the drone operators, live in civilian neighborhoods and never need to deploy from garrison to accomplish their missions. Some service members spend most of their careers working almost exclusively for civilians. Some have seen their duties turned over to contractors who are now responsible for many of the vital, dangerous tasks once done by those in uniform. Some in uniform have learned elite skills and joined specialized organizations; although they continue to wear the uniform of one of the four services, they have little connection to it. And some military members, especially those responsible for land-based nuclear weapons, aspire to an ironic measure of success: *not* performing the tasks they train for, against a threat that the nation has largely forgotten since the end of the Cold War.

The evolving organizational relationships exacerbate this blurring of purpose for individuals. The growing ascendency of the combatant commands (COCOMs) under the 1986 Goldwater-Nichols Act disrupted the services' traditional roles.[1] Although COCOMs were originally intended to be headquarters to conduct current operations, they have gradually evolved into organizations that not only

[1] Public Law 99-443, Goldwater-Nichols Depart of Defense Reorganization Act of 1986, October 1, 1986.

compete with services but often supersede them. Since the passage of Goldwater-Nichols, COCOM commanders have had direct access to the Secretary of Defense, which is not the case for the service leaders. The legislation essentially subordinated the military services to the COCOMs, turning the former into force providers. With this change, only the COCOMs fight, in theory. General Tommy Franks, as the commander of U.S. Central Command (CENTCOM), graphically underscored this point with his scathing reference in *American Soldier* to the service chiefs as "Title Ten mother-f------" when they commented on his war plan in their Joint Chiefs of Staff (JCS) capacity. From his perspective, the JCS had little right to comment on his war plan—rather, the JCS existed to provide the traditional "bullets and beans" to support the plan.[2] The military services, once revered and organizationally supreme, now assume second-tier status to that of the COCOMs.

Even more ironic, in the intervening years since 1986, even the geographic COCOMs have become more focused on developing proxy-fighting capabilities through cooperation and building partnership capacity rather than combat. At least in the case of the last two wars, in Afghanistan and Iraq, separate operational commands under CENTCOM conducted the actual operations, while the commander assumed a quasi-diplomatic role. So, despite legislative efforts to clarify responsibilities, in a practical sense, the person or organization ultimately held responsible for the success or failure of a combat operation is difficult, if not impossible, to determine.

Further complicating this relationship are the chameleonlike organizations that have emerged as a result of the Goldwater-Nichols Act and other legislation. U.S. Special Operations Command (SOCOM), for example, is a specialized COCOM with the legal attributes of the military services yet frequently overlapping in responsibilities with the Central Intelligence Agency (CIA).[3] U.S. Cyber Command, on the

---

[2] Tommy Franks, *American Soldier*, New York: Regan Books, 2004, pp. 207, 277.

[3] Depending on its mission, SOCOM, in particular, can operate under either Title 10 of the United States Code governing U.S. Department of Defense and military operations, or Title 50, governing intelligence activities and covert action. For discussion of this debate, see

other hand, arguably relies on a heavily civilianized organization that acts with great autonomy while, in theory, being subordinate to a functional COCOM, U.S. Strategic Command.[4]

This is just a sampling of the issues associated with America's 21st-century military. Individually, each can be explained, but, collectively, the explanations seem overly involved, complicated, and ultimately not compelling. What follows is an attempt to use this colonel's questions to initiate a larger conversation about the trajectory of America's 21st-century military.

Organized Violence: The Military's Unique Responsibility

As the colonel had reflected, three qualities seemingly made members of America's military stand apart from their civilian counterparts. First, members of the military take an oath to support and defend the Constitution; second, they wear a uniform that identifies them with a specific service; and, third, they subject themselves to a specific legal code, the Uniformed Code of Military Justice, that exerts considerable control over personal and professional behavior. However, a quick examination reveals these qualities are by no means unique to the military. Even though they set military members apart from many private-sector civilians and signify a position of public trust, they are also qualities shared with millions of others outside the military.

Andru E. Wall, "Demystifying the Title 10-Title 50 Debate: Distinguishing Military Operations, Intelligence Activities and Covert Action," *Harvard National Security Journal*, Vol. 3, 2011.

[4] See, for example, discussions of changes in the intelligence and cyber areas in Dana Priest and William Arkin, "Top Secret America," *Washington Post*, July 19–20, 2010; Ben FitzGerald and Parker Wright, *Digital Theaters*, Center for New American Security, April 2014. Note that the dual-hatted Commander of Cyber Command and Director of the National Security Agency (NSA) has unique responsibilities under both Title 10 and Title 50; see Secretary of Defense, "Establishment of a Subordinate Unified Cyber Command Under U.S. Strategic Command for Military Cyberspace Operations," memo, Washington, D.C., June 23, 2009; Robert Gates, *Duty: Memoirs of a Secretary at War*, New York: Alfred A. Knopf, 2014, Chapter 12.

For example, the oath of office is far from exclusive to the military. Article VI of the Constitution requires that senators and representatives, "the Members of the several State Legislatures, and all executive and judicial Officers, both of the United States and of the several States, shall be bound by Oath or Affirmation, to support this Constitution."[5] In this spirit, the military officer and civilian service oaths are virtually identical. Office holders swear to "support and defend the Constitution of the United States against all enemies foreign and domestic." The donning of a uniform is similarly identified with public trust and suggests to others that the individual wearing the uniform has made a personal sacrifice to fulfill this responsibility. But, again, the military is not alone. Most notably, police officers and firefighters also make sacrifices in fulfilling their public roles, at the state and local levels. Finally, submitting to the Uniformed Code of Military Justice acknowledges a willingness to be held to a higher standard than some people in the private sector, but many in the government and private-sector positions similarly subject themselves to higher standards, whether they are governed by professional codes or fiduciary responsibilities. The oath, uniform, and code are important preconditions, but they certainly do not set military members apart from those who serve in other public institutions.[6]

What then makes the armed forces unique in a nation such as the United States? Fundamentally, that quality is the responsibility for managing and employing organized violence on behalf of the nation. Importantly, the organized violence controlled by the military is not targeted against individuals. As Hedley Bull observed in *The Anarchical Society*,

> War is organised violence carried out by political units against each other. Violence is not war unless it is carried out in the name of a political unit; what distinguishes killing in war from murder

---

[5] The text of Article VI is available on the website of the National Archives (see http://www.archives.gov/exhibits/charters/constitution_transcript.html).

[6] Military personnel do make a unique commitment by signing up for two to six years of service, which they cannot unilaterally cut short and could theoretically take them anywhere to conduct any lawful mission.

> is its vicarious and official character, the symbolic responsibility of the unit whose agent the killer is. Equally, violence carried out in the name of a political unit is not war unless it is directed against another political unit; the violence employed by the state in the execution of criminals or the suppression of pirates does not qualify because it is directed against individuals.[7]

The military is America's means to employ violence against other political units in order to address national problems.

Two other scholars who have studied organized violence explained why this is important. Mary Kaldor in *New and Old Wars* noted that recognizing the importance of socially organized, legitimate violence is equally important to the unit and to the individual soldier, giving the latter a common goal in which to believe and to share with others.[8] This belief, according to Dave Grossman in *On Killing*, gives those responsible for organized violence understanding and affirmation. Speaking specifically about those in the U.S. military, Grossman observed that it was important for American veterans to understand "that they did no more and no less than their nation and their society asked them to do; no more and no less than American veterans had honorably done for more than two centuries. And . . . that they [were] good human beings."[9] To give meaning to their service, those in America's military must see their role in organized violence as important to the nation.

Thus, the U.S. military, like other national militaries, assumes a responsibility to provide organized-violence solutions to national problems. How nations fill the military ranks and organize for this purpose varies widely depending on variety of factors, including geography, type of government, culture, history, economic strength, population, and the unique problems facing a nation. But, in all cases, providing

---

[7] Hedley Bull, *The Anarchical Society: A Study of Order in World Politics*, New York: Columbia University Press, 1995, p. 178.

[8] Mary Kaldor, *New and Old Wars: Organized Violence in a Global Era*, 3rd ed., Cambridge, UK: Polity Press, 2012, p. 28.

[9] Dave Grossman, *On Killing: The Psychological Cost of Learning to Kill in War and Society*, New York: Back Bay Books, 2009, p. 299.

organized-violence solutions to national problems underpins the existence of national militaries.

Thomas Kuhn on Problems, Paradigms, Anomalies, and Paradigm Shifts

A discussion of national problems and their possible solutions leads quite naturally to Thomas Kuhn's *The Structure of Scientific Revolutions* and his insights on problems, paradigms, anomalies, and paradigm shifts. Kuhn, a Harvard-trained physicist, became one of the 20th century's most eminent historians and philosophers of science. He wrote his work in an effort to understand the nature of scientific advancements. He wondered whether scientists had been fundamentally misled about their discipline because the history of science was taught as a linear process characterized by "the accumulation of individual discoveries and inventions."[10] He posited instead that most science, or "normal science," as he called it, "is predicated on the assumption that the scientific community knows what the world is like."[11] That worldview subsequently provided an accepted model, or paradigm, through which other problems can be solved. Kuhn pointed out that an established paradigm is highly cumulative, is successful in its aim, and extends the body of knowledge. It does not look for novelties of fact and theory, and, at its most successful, uncovers none. A strong paradigm drives the construction of elaborate equipment and carries with it its own language, rules, and standards for behavior and practice. Finally, it establishes the criteria for choosing problems in the first place and then provides the means to solve acute problems, or anomalies, facing a group of practitioners. This means that, when the

---

[10] Thomas S. Kuhn, *The Structure of Scientific Revolutions*, Chicago: University of Chicago Press, 2012, pp. 1–2. See David E. Johnson, *Modern U.S. Civil-Military Relations: Wielding the Terrible Swift Sword*, Washington, D.C.: National Defense University Press, 1997, for another application of Kuhn's work.

[11] Kuhn, 2012, p. 5.

paradigm itself is taken as a given, any problem associated with it is assumed to be soluble or at least explainable.[12]

A compelling paradigm can cope with myriad anomalies by solving them, ignoring them as irrelevant, or explaining them away. Over time, however, even the strongest paradigms fail to solve extraordinary problems, creating a crisis that loosens the rules for normal problem-solving and creating the space for alternatives to be considered. As a new paradigm emerges, incommensurability with its predecessor is inevitable.[13] In *The Structure of Scientific Revolutions*, Kuhn explained incommensurability by describing how a scientific community committed to a Ptolemaic view of the universe (where the sun revolved around the earth) faced great dissonance when presented with Copernicus's evidence that the earth in fact revolved around the sun. Those adhering to the Ptolemaic paradigm faced an incommensurable problem—they could hold on to their existing worldview by challenging Copernican assumptions, problem categorizations, and evidence, or they could go through a shift, a conversion process, accepting Copernicus's evidence and dramatically altering their intellectual and personal perspectives. Almost all chose the former course; societal pressures, their intellectual identities, and existing frameworks were too strong to accept the change, assuming that they even recognized that there was an issue.[14]

This example also highlights another key aspect of Kuhn's thinking on incommensurability. The emergence of a competing paradigm does not mean that everything from the earlier paradigm is necessarily cast aside; rather, it fundamentally means that, with different criteria for identifying and solving problems, "the proponents of competing paradigms practice their trades in different worlds."[15] During the period when both paradigms compete for primacy, one can choose to live in a Ptolemaic world or a Copernican world, but the differences are so stark that one cannot live in both. Most important, every-

[12] Kuhn, 2012, pp. 5, 23–24, 37, 52–53, 64–65, 76. Kuhn specifically refers to the solving of other problems within the paradigm as "normal science."

[13] Kuhn, 2012, pp. 76, 80.

[14] Kuhn, 2012, pp. 68–70, 147.

[15] Kuhn, 2012, p. 149.

one must choose. Not choosing a paradigm is a choice as well. Tacitly accepting the discontinuities in one paradigm while recognizing the emergence of another paradigm means that even recognized problems can no longer be solved. They can only be temporarily addressed with decreased effectiveness and increased complexity.

Kuhn's questioning of the traditional nature of scientific advancement seized the imagination of scholars throughout the academic world and inspired them to look anew at their disciplines.[16] Such a problem-paradigm approach could be similarly useful for better understanding the U.S. military and more clearly illuminating its mission and contemporary challenges. The Constitution's preamble requirement "to provide for the Common Defense" provides the shorthand mission statement for America's military. Although these six words are a brief statement, they encapsulate a dynamic, ongoing national debate over what constitutes the common defense and how the United States understands the role of organized violence in providing for it.[17]

Two distinct historical paradigms—the continental and overseas—stand out when using Kuhn's problem-paradigm approach to explain how the United States provided for the common defense. They also suggest that America's contemporary understanding of the common defense has reached a Kuhnian crisis point from which a third paradigm has emerged. Four aspects shared by both paradigms reveal what is central to them and why the shift occurred: (1) *location*, specifically, the predominant geographic area that warranted a common defense; (2) *concept of violence*, specifically, the type of organized violence associated with the common defense; (3) *organization and people*, specifically, the combination of people, organizations, and structures that bore responsibility for mastering and employing organized violence;

---

[16] Alexander Bird, "Thomas Kuhn," revised August 11, 2011, in *The Stanford Encyclopedia of Philosophy*, Stanford, Calif.: Metaphysics Research Lab, Center for the Study of Language and Information, Stanford University, 2013.

[17] Several conversations during my military career inform this view. When I asked service members what the U.S. military did, they routinely replied, "Support and defend the Constitution." When asked to explain what that meant, they answered, "Provide for the common defense." Finally, I asked, "What is the common defense?" The answer invariably was some version of, "Whatever the nation's leaders say it is."

and (4) *organizational culture*, specifically, the formal and informal cultures that developed within these organizations and structures. A fifth aspect, *incommensurability*, identifies when a paradigm is in crisis and foreshadows its shift. Although the majority of this exploration focuses on the overseas paradigm, understanding the original common defense problem-paradigm pairing that emerged after the American Revolution highlights the importance of this enduring dynamic.[18]

The Common Defense and the Continental Paradigm

The continental paradigm reflected the unique circumstances facing the newly constituted United States of America. Geographically isolated from their enemies, short on resources, and mistrusting of standing armies, the framers of the Constitution debated at length the meaning of *the common defense*, what type of military that implied, and why the United States needed that military to solve the challenges the new nation faced. The debate between the Federalists and the Anti-Federalists framed the argument. Federalists, such as Alexander Hamilton, highlighted the dangers posed by the expanded presence of Britain and Spain in North America; hostilities from Native populations on the frontier, abetted by Britain and Spain; and the inability to secure the Atlantic seaboard to protect commercial interests. Hamilton argued that, collectively, these dangers encircled the new nation and were therefore common to all the states, and the essence of the common defense was to address these dangers. To do so successfully meant that the federal government, not the individual states, needed to be given the authority for "the formation, direction or support of the national forces" essential to the common defense.[19] While the Anti-Federalists

[18] Mark Grimsley, "Master Narrative of the American Military Experience: The American Military History Narrative: Three Textbooks on the American Military Experience," *The Journal of Military History*, July 2015, pp. 798–802, gives a good overview of the common narratives associated with the evolution of America's military.

[19] Alexander Hamilton, "The Necessity of a Government as Energetic as the One Proposed to the Preservation of the Union," Federalist Paper No. 23, Library of Congress, 1787; Alexander Hamilton, "The Powers Necessary to the Common Defense Further Considered,"

disagreed with Hamilton's characterization on almost every point, their inability to present persuasive counterarguments and the tangible shortcomings of the Articles of Confederation meant that the Federalist argument largely carried the day, as reflected in the Constitution.[20] Thus, this first paradigm, which provided for the common defense, assumed the following characteristics.

Location. While the Anti-Federalists saw a geographic location limited to the boundaries of the new states and little need to be involved at sea, the Federalists conceived a much larger geographic area in need of protection. The Federalists looked beyond existing borders and allowed for their possible expansion across the North American continent. The Federalists also sought to protect and expand seaborne trade interests as much as possible, recognizing that the United States could not compete with the great seafaring nations of the day. The Federalists' more expansive location for the common defense won out and, for the next century, shaped how the United States would think about where its military operated and what it should do.

Concept of violence. Individual experiences from the American Revolution shaped contemporary understanding of organized violence as something that was personal, proximate, reciprocal, and potentially fatal. In other words, as instruments of organized violence, those who served in the American Revolution assumed that they were similarly at risk of being subject to violent actions. This would have echoed Carl von Clausewitz's observation a few decades later that understanding war required understanding the danger associated with this act of organized violence. Clausewitz explained this danger by describing a battlefield filled with "cannonballs," "bursting shells," shots "falling like

Federalist Paper No. 24, Library of Congress, 1787; Alexander Hamilton, "The Same Subject Continued: The Powers Necessary to the Common Defense Further Considered," Federalist Paper No. 25, Library of Congress, 1787.

[20] Brutus, "Certain Powers Necessary for the Common Defense, Can and Should Be Limited," Antifederalist Paper No. 23, The Federalist Papers Project, 1788a; Brutus, "Objections to a Standing Army (Part I)," Antifederalist Paper No. 24, The Federalist Papers Project, 1788b; Brutus, "Objections to a Standing Army (Part II)," Antifederalist Paper No. 24, The Federalist Papers Project, 1788c; and Richard Kohn, *Eagle and Sword*, New York: Free Press, 1975, Chapter 5.

hail," "hissing bullets," and "the sight of men being killed and mutilated." Coping with such an environment, much less operating effectively in it, Clausewitz observed, required those involved to accept as routine tremendous physical exertion, privation, physical discomfort, and fatigue.[21] The Continental Army veterans of Trenton, Saratoga, and Valley Forge or those serving on John Paul Jones's USS *Bonhomme Richard* would have related to this portrayal and described their respective experiences along similar lines.

Organization and people. Similarly, how the framers understood violent means, organized them, and then used them to seek solutions to problems was consistent with the era. The new U.S. military's organization, unsurprisingly, was based on physical geography. This organization provided for a land and a sea force. The role of solutions based on organized violence, in light of these external threats, was formally codified in the Constitution. Article I, Section 8, established the right to raise and support an army, provide and maintain a navy (including a small portion of seaborne infantry or marines), and provide for calling forth a militia. The relative size of the military, especially compared with militaries in Europe, was small, and volunteers provided the manning. Since the U.S. military during this period lacked any common approach to training, the relative success of the mastery of violent means rested with individual garrison or ship commanders.

Organizational culture. Although life in an isolated frontier garrison and aboard a ship would bear little resemblance to each other at first glance, the similarities of the day-to-day isolation, physical hardship, and routine danger associated with merely surviving in these two environments meant that the essence of their organizational cultures was quite similar. To cope with these hardships, martial qualities were essential. For example, group needs in garrison or aboard ship dominated individual needs, all military members operated in a hierarchy, with an understanding of each individual's role in that hierarchy, and

---

[21] Carl von Clausewitz, *On War*, ed. and trans. Michael Howard and Peter Paret, Princeton, N.J.: Princeton University Press, 1984, pp. 113–116.

unit order allowed for tough disciplinary means routinely employed by the commander.[22]

The continental paradigm, in short, focused on the North American landmass and key seafaring trade routes. Its concept of violence was both personal and proximate. Those serving on land or at sea lived in a more inherently dangerous environment and operated with a concept of violence that was personal, proximate, and reciprocal. The individuals responsible for organized violence were volunteers organized based on physical geography into an army and a navy. Finally, because of the small size of these organizations, relative to the large areas to be covered, the organizational cultures, while martial in nature, were also greatly dependent on individual commanders.

This continental paradigm dominated throughout the 19th century. In America's first century after the Constitution's ratification, the use of the U.S. military to provide violence-based solutions remained largely land focused, but the sea component played the important trade role the framers envisioned. This use of organized violence included the War of 1812, the Mexican War, and, most notably, the American Civil War. Even the American Civil War, despite its size and scope, was ultimately consistent with the continental paradigm. The war was the ultimate effort to provide for the common defense. It was nothing less than a struggle for the maintenance of the Union and national borders. For the Union, the war demanded a large American land army to defeat the breakaway states and a navy sufficiently large to blockade Confederate supply lines. Throughout the conflict, the dominant location remained the continent; the concept of violence was personal, both in terms of inflicting violence and being subjected to it; the organization was based on physical geography; and the organizational cultures were unit focused, hierarchical, and disciplined.

Moreover, one of the most pronounced anomalies, the implementation of an unpopular draft in 1863 to fill Union ranks, was accommodated by the continental paradigm. President Abraham Lincoln and Congress invoked their authority under the Constitution

[22] Kohn, 1975, Chapter 14; and Edward M. Coffman, *The Old Army: A Portrait of the American Army in Peacetime, 1784–1898*, Oxford, UK: Oxford University Press, 1986, Chapter 4.

"to suppress insurrection and rebellion," since "no service can be more praiseworthy or honorable than that which is rendered for the maintenance of the Constitution and the Union." As the Enrollment Act (the "Act for enrolling and calling out the national Forces") further noted, to achieve this goal "a military force is indispensable," thus allowing the President to call up male citizens between the ages of 20 and 45 to serve.[23] The act directed draftee call-up by state districts and focused on raising sufficient manpower to round out individual units for the duration of the crisis. This act was controversial at the time. It was full of exceptions, riddled with problems, and, most notably, prompted the New York draft riots of 1863. But for all its shortcomings, it did contribute to a stronger sense of national citizenship than what existed before the war. Not surprisingly, as soon as the Civil War ended, the draft was quickly terminated.[24]

Once the American Civil War ended, the Army and Navy rapidly reduced in size and returned to their more traditional roles protecting, expanding, and eventually closing the land frontier, as well as patrolling and securing trade routes. The continental paradigm was firmly in place. Indeed, in 1898, on the cusp of war with Spain, the Regular Army numbered a mere 28,000—the same size it had been since the 1870s. And the Navy possessed an obsolete hodgepodge of ships in various states of disrepair and was only beginning to modernize this obsolete post–Civil War fleet as the end of the 19th century approached.[25]

Incommensurability. The continental paradigm appeared sound even in the last decade of the 1800s. During the period from the Spanish-American War to U.S. entry into the First World War, however, the continental paradigm abruptly experienced a critical anom-

---

[23] U.S. Congress, 37th Cong., 3rd Sess. Chapters 74 and 75, Enrollment Act, Washington, D.C., March 3, 1863.

[24] James McPherson, *Battle Cry Freedom: The Civil War Era*, Oxford, UK: Oxford University Press, 1988, pp. 600–611; John Whiteclay Chambers II, *To Raise an Army: The Draft Comes to Modern America*, New York: The Free Press, 1987.

[25] Edward M. Coffman, *The Regulars: The American Army, 1898–1941*, Cambridge, Mass.: Harvard University Press, 2007, p. 3; and Peter Karsten, *Naval Aristocracy: The Golden Age of Annapolis and the Emergence of Modern American Navalism*, Annapolis, Md.: Naval Institute Press, 2008, pp. 278, 288, 307.

aly that it could not accommodate. While the concepts of violence, organizations, and organizational cultures were largely consistent with the continental paradigm, there was a remarkable shift in the dominant location for the paradigm. After the Spanish-American War, the majority of American soldiers ended up stationed overseas for the first time, especially in the Philippines. The Navy, meanwhile, both advocated for and happily embraced the global mission highlighted by its key role in the Spanish-American War. Congress's willingness to fund the Great White Fleet, combined with President Theodore Roosevelt's decision to deploy it on a global goodwill mission, in December 1907, demonstrated the degree to which the overseas dimension of the common defense suddenly dominated.[26] As Kuhn might note, these large anomalies suggested a looming paradigmatic crisis in how to think about the common defense: Was it focused on the North American continent, or, in fact, had the nation's understanding of the geographic epicenter of the common defense shifted? With America's entry into the First World War, that question was emphatically answered.

The Common Defense and the Overseas Paradigm

On April 2, 1917, President Woodrow Wilson stood before the U.S. Congress and declared that the German Navy's sinking of American ships and taking of American lives constituted "a warfare against mankind. It is a war against all nations." Wilson went on to note that

> [war] will involve the organization and mobilization of all the material resources of the country to supply the materials of war and serve the incidental needs of the nation in the most abundant and yet the most economical and efficient way possible. It will involve the immediate full equipment of the Navy in all respects but particularly in supplying it with the best means of dealing with the enemy's submarines. It will involve the immediate addi-

---

[26] Ian Toll, *Pacific Crucible: War at Sea in the Pacific, 1941–1942*, New York: W.W. Norton and Co., 2012, Prologue.

> tion to the armed forces of the United States already provided for by law in case of war at least 500,000 men, who should, in my opinion, be chosen upon the principle of universal liability to service, and also the authorization of subsequent additional increments of equal force so soon as they may be needed and can be handled in training.[27]

With this speech, Wilson challenged the nation to undertake a task requiring organized violence unlike any it had ever encountered and on a scope it had never previously envisioned.

Location. Wilson's recommendation to take the nation to war to fight "for the rights of nations great and small and the privilege of men everywhere to choose their way of life and of obedience" created a crisis for the continental paradigm that it could not accommodate.[28] Wilson's declaration and commitment to cooperate with allies fighting Germany in Europe confirmed the geographic shift overseas that had first started two decades earlier, with the Spanish-American War. Indeed, when the United States entered the war in April 1917, it faced a new, enormous problem: What was the best way for the United States to create and move a large national military overseas into an extremely violent environment? The continental paradigm was insufficient to explain the geographic shift in this understanding of the common defense, much less how to conceptualize and create the force responsible for organized violence in this environment. Thus, with Wilson's speech, the United States left the North American continent behind as the dominant geographic location for the common defense and moved it overseas. Geographic and technological limitations meant that, if the United States wanted to engage Germany, it had to cross the Atlantic Ocean to do so.

Concept of violence. Based on the horrific pictures and accounts of such battles as Verdun and the Somme, America's leaders assumed that those serving on the Western Front, or flying over it, would experience

[27] Woodrow Wilson, "Wilson's War Message to Congress," World War I Document Archive, April 2, 1917.

[28] Wilson, 1917.

a personal proximity to organized violence, but this violence would reflect the brutal industrialized nature of the era, whether fighting in the trenches or engaging in early aerial combat over them. Similarly, whether on convoy duty or enforcing the blockade of Germany in the North Atlantic, American naval personnel would face the routine challenges of operating at sea, as well as the new uncertainties and potential dangers of submarine warfare. Thus, the First World War reinforced for America's soldiers, sailors, and marines a shared sense of danger and personal relationship to organized violence, even while they engaged in a new type of industrialized warfare.

Organization and people. To manage organized violence on such a large scale, America's military was structured in a manner consistent with the other nation-states of the era and was supported by large-scale industrial mobilization. This required the nation to consider, for the first time, how it would raise and maintain a large national army overseas, including a nascent army air corps and a maritime force capable of commanding the seas. This also required judgments about which overseas territories or waters needed protection or occupation. Physical geographic spaces therefore continued to define military organizations that, because of technological limitations, if nothing else, still operated autonomously from one another.[29] However, these spaces were increasingly distant from America's shores. On land and at sea, at the highest organizational levels, the Army and Navy continued to operate separately, and their uniformed commanders were held accountable for identifying and achieving America's wartime objectives in their geographic areas.[30]

Once the geographic shift was made and the organizational structure validated, the nation had to address the huge manpower demands that accompanied America's commitment to fight in Europe. This led Congress to authorize, and President Wilson to approve, the total con-

[29] Today, the U.S. military refers to this as *domains*; they include land, sea, and, starting in the First World War, air.

[30] Frank E. Vandiver, "Commander-in-Chief—Commander Relationships: Wilson and Pershing," *The Rice University Studies*, Vol. 57, No. 1, 1971. While Pershing led the American Expeditionary Forces, MG Tasker Bliss served as Army Chief of Staff and Secretary of War Newton Baker's closest advisor for most of the First World War.

scription of "all male persons between the ages of 21 and 30." In his proclamation on the draft to the American people, Wilson went on to state:

> The whole nation must be a team, in which each man shall play the part for which he is best fitted. To this end, Congress has provided that the nation shall be organized for war by selection; that each man shall be classified for service in the place to which it shall best serve the general good to call him.
>
> The significance of this cannot be overstated. It is a new thing in our history and a landmark in our progress. It is a new manner of accepting and vitalizing our duty to give ourselves with thoughtful devotion to the common purpose of us all.
>
> It is in no sense a conscription of the unwilling; it is, rather, selection from a nation which has volunteered in mass. It is no more a choosing of those who shall march with the colours than it is a selection of those who shall serve an equally necessary and devoted purpose in the industries that lie behind the battle line.
>
> The day here named is the time upon which all shall present themselves for assignment to their tasks. It is for that reason destined to be remembered as one of the most conspicuous moments in our history. It is nothing less than the day upon which the manhood of the country shall step forward in one solid rank in defence of the ideals to which this nation is consecrated.
>
> It is important to those ideals no less than to the pride of this generation in manifesting its devotion to them, that there be no gaps in the ranks.[31]

Antiwar activists hoped for large protests against this new draft, similar to those in the American Civil War. They were disappointed, however, in the reactions. The American people and the U.S. government instead acknowledged a shared responsibility that allowed citi-

---

[31] Woodrow Wilson, "President Wilson's Proclamation Establishing Conscription," Firstworldwar.com, May 28, 1917.

zens to be drafted into the military when the U.S. government deemed that there was imminent risk to the nation. Conscription thus illuminated a new reciprocal relationship between the U.S. government and its population.[32] The American people looked to the government to provide for the nation's defense, and in return the government expected its citizens to serve if it perceived a threat to the nation.

Organizational culture. Moreover, whatever their unique attributes, the institutional cultures of the military services continued to share a common objective and responsibility: to master, manage, and employ organized violence effectively. How the services understood this violence varied based on their physical environments. The Marine Corps in particular had to adjust in a significant way as it came ashore for the first time and integrated into the American Expeditionary Forces (AEF). But, during the First World War, America's Army, Navy, and Marine Corps focused on ensuring that their service members, whether volunteers or conscripts, were personally connected to an aspect of organized violence.[33] As a result, these American service members shared a cultural core based on the need to be physically prepared for the demands and personal risk associated with organized violence. The personal risk inherent in organized violence heightened the need for a standardized organizational approach to basic training. Unlike the 19th-century unit-centered military training, this initial training placed a service-wide emphasis on fostering martial qualities, such as commitment to rigid hierarchy, strict unit and self-discipline, athleticism, and physical courage under fire. The more difficult the circumstances—on land,

---

[32] See Chambers, 1987, especially Chapters 7–8, on the implementation of and reaction to the draft, in 1917.

[33] Even the Marine Corps Fourth Infantry Brigade that came ashore and integrated into the 2nd U.S. Army Division of the AEF managed to retain a unique service identity. According to Marine Corps lore, this brigade performed with such great distinction at Belleau Wood that it gave rise to the Marine Corps nickname "devil dogs." See Edwin McClellan, *United States Marine Corps in the World War*, Washington, D.C.: Headquarters, U.S. Marine Corps, especially Chapters 9–10; Richard W. Stewart, ed., *The United States Army in a Global Era, 1917–2008*, Vol. 2, *American Military History*, Washington, D.C.: Center of Military History, 2010, pp. 6–52.

at sea, or in the air—the more important these characteristics were to effectively wielding organized violence, much less individual survival.[34]

Even in the chaos of America's early months of mobilization, this training started immediately when new recruits entered basic training. All were introduced to the physical demands of military life, with an emphasis on martial skills, physical fitness, operating together in confined spaces (especially at sea), and the importance of wielding violence effectively even in the face of mortal danger. Service members' wartime experiences on land and at sea tied directly to the experience or expectation of personally facing organized violence.[35] In America's nationally conscripted military, wherever those in uniform served, whether in Western Europe or the Atlantic, organized violence and its associated personal risks provided the dominant cultural component.[36]

Collectively, these attributes were America's answer to the vexing question of how to create, move, and use a large military overseas during the First World War. Indeed, the United States and its military had shifted to a new paradigm in the process of confronting the incommensurable problem of fighting Germany overseas. Despite considerable continuity in the understanding of violence, even with its industrialized nature, as well as organizational structure and cultures, the century-old continental paradigm looked problematic in the face of huge manpower requirements. The paradigm failed when the dominant location of the common defense shifted decisively overseas. To cope, the United States needed to create a new, compelling solution

---

[34] Chambers, 1987, pp. 144–149, 196–197; Coffman, 2007, Chapter 4.

[35] Martial skills training in hand-to-hand combat can be viewed at Army VideoTube, "U.S. Army Hand to Hand Combat Training," video posted to YouTube, April 9, 2013; see also Doughboy Center, "In Their Own Words: The Story of AEF by Its Members, Allies and Opponents in Seven Parts," *The Story of the American Expeditionary Forces*, Worldwar1.com, undated.

[36] See Michael Howard, *War in European History*, Oxford, UK: Oxford University Press, 1984, pp. 109–110; John Hackett, *The Profession of Arms*, New York: Macmillan Publishing Co., 1983, p. 141; James J. Sheehan, *Where Have All the Soldiers Gone? The Transformation of Modern Europe*, New York: Mariner Books, 2009, pp. 21, 38; Samuel Huntington, *The Soldier and the State: The Theory and Politics of Civil-Military Relations*, Cambridge, Mass.: Harvard University Press, 1985, pp. 7–18.

or paradigm for large-scale, overseas, nation-state conflict. Consistent with Kuhn, this new overseas paradigm also possessed its own rules and standards for behavior and practice and, most important, provided the means to once again solve acute problems, or anomalies, facing a group of practitioners.[37]

The Second World War Validates the Overseas Paradigm

Immediately after the November 11, 1918, armistice, the draft ended, the American military quickly demobilized several million men, and the nation soon retrenched. The Army saw its numbers contract more than tenfold, but, consistent with the new paradigm, those who did remain were mostly stationed overseas. Relative to the Army, the Navy was in a stronger position. But the naval-limitation treaties and the demobilization after the First World War took a toll as well. Those few remaining in uniform practiced their craft on land from geographically isolated garrisons at home or overseas; and at sea, they operated far from shore for extended periods. The First World War had added the atmosphere above the battlefield as yet another place to manage and employ organized violence. But the practitioners of this type of organized violence operated in similarly isolated geographic environments. In short, those in uniform lived and operated largely separate from American civil society, whether overseas or in the United States.[38]

Location. However, the planning horizon and expectations for future conflicts clearly remained overseas. Through the 1920s and much of the 1930s, Army and Navy planners focused on developing the color plans for overseas operations. Most notably, War Plan Orange anticipated a future conflict with Japan, and Red-Orange planned for a British-Japanese alliance. The former anticipated a major conflict with Japan in the Pacific. The latter, though considered unlikely, placed the

---

37 Kuhn, 2012, pp. 23–34.

38 There are certainly some notable exceptions, such as the court-martial of Billy Mitchell, the Army's role in dispersing the Bonus Marchers, and the close ties between the Army and the Civilian Conservation Corps.

major threat in North America, the Caribbean, and Latin America. By late 1937, the Anti-Comintern Pact uniting Germany, Italy, and Japan against communism again changed the strategic landscape for America's war planners, this time by making the likelihood of a two-front war much greater. The new Rainbow plans evaluated a variety of planning challenges associated with this three-way partnership, but the overseas focus remained constant. Most presciently, this effort produced the Rainbow 5 plan, which put the United States in a defensive coalition in the Pacific while rapidly projecting power across the Atlantic, into Western Europe.[39]

In addition to more-robust planning, based on German and Japanese expansionistic moves, President Franklin Roosevelt moved steadily toward national mobilization in the late 1930s. Almost a year before Pearl Harbor, Roosevelt and his advisors endorsed most of Rainbow 5, including the defeat of Germany and Italy as America's top priority. Once that was accomplished, the nation would deal with Japan.[40] After the attack on Pearl Harbor, the entire nation went to war using the overseas paradigm, as outlined in Rainbow 5 and subsequent plans, as the blueprint for war in the Atlantic and the Pacific. Ultimately, the Second World War proved the culmination of the nation-state era of overseas conflict on a global—indeed, epic—scale.

Concept of violence. In keeping with the overseas paradigm, the personal connection to organized violence remained largely the same during the Second World War. Despite the massive industrial effort that underpinned America's fighting ability, those who joined the military still assumed that they were at greater personal risk than civilians at home and that, by joining the military, they were assuming a close, proximate relationship to violence that increased the likelihood of killing and being killed. This expectation had been set by the vast scope of organized violence in Europe and Asia prior to America's entry and was immediately and dramatically reinforced by the simultaneous attacks on Pearl Harbor, the Philippines, and Wake Island. Over the course of

---

[39] Kent Greenfield, ed., *Command Decisions*, Center of Military History, United States Army, Washington D.C.: Government Printing Office, 1958, pp. 21–44.

[40] Greenfield, 1958, p. 42.

the next four years, U.S. military campaigns required the management and employment of violence on a massive scale, across vast regions in the Atlantic and the Pacific. They resulted in more than 420,000 American deaths, and many more were wounded. This number is small when compared with the almost–20 million total battle deaths from the Second World War, but, for the United States, only the American Civil War resulted in more deaths.[41]

Organization and people. Consistent with the overseas paradigm, autonomous military services and operations continued to dominate the American approach to the war; however, the creation of an informal JCS and the identification of a senior military advisor to the President nudged the nation to more interservice cooperation at the theater level, especially in Europe. From General Dwight Eisenhower on down, the role of the commander and his associated responsibilities were apparent at every level of command. However, when it came to executing the actual operations, the commanders continued to rely, for myriad reasons, on the services fighting within their geographic spaces, as they had in the First World War.[42]

The relationship between the U.S. government and the American people also remained reciprocal and heavily reliant on conscription. From a personnel perspective, the nation ultimately mobilized more than 12 million Americans. The interwar Army grew to a force of approximately 6 million, and the Navy swelled to 3.4 million. The magnitude of the conflict similarly turned the Army Air Forces into a large military service of more than 2 million, focused on strategic bombing and with nominal ties to the Army as the war progressed. Finally, the amphibious operations against enemy-fortified islands, which characterized much of the war in the Pacific theater, resulted

---

[41] The number of deaths in the war remains difficult to estimate. The National World War II Museum offers the following estimates: 15 million battle deaths and 45 million civilian deaths. See National World War II Museum, "By the Numbers: World-Wide Deaths," web page, undated-b.

[42] This insistence on autonomy was a source of friction between General Douglas MacArthur and Admiral Chester Nimitz in the Pacific theater and ultimately became a challenge even for the nation's most-senior leaders; see Steven L. Rearden, *Council of War*, Washington D.C.: National Defense University Press, 2012, pp. 29–33.

in the size of the Marine Corps increasing from fewer than 20,000 marines in 1939 to more than 475,000 by the end of the war.[43]

Organizational culture. All the services continued to foster organizational cultures that emphasized the importance of hierarchy in managing and employing their particular version of organized violence. Whether on land, at sea, or in the air, this included the expectation of prolonged absence from home and the likelihood of encountering great personal, even fatal, risks while in uniform. The service-wide approach to training introduced when the United States entered the First World War was refined, although far from perfect, when the nation entered the Second World War. Each military service had endeavored to standardize its training approach in a way that matched its unique responsibilities. But all services focused on instilling martial qualities—such as courage under fire, physical stamina in extremely demanding circumstances, and creating a culture that stressed survival, much less national victory—depended on putting the unit's success above all else, especially individual concerns and needs.[44]

Thus, the U.S. military entered and subsequently emerged from this conflagration with the overseas paradigm fully intact. Indeed, several organizational-culture characteristics of the overseas paradigm—including the ability to sacrifice unto death for the nation and to cope with great hardship, personal risk, and loss—not only endured but were revered. Not surprisingly, then, the overseas paradigm continued to provide the context for the nation's civilian and military leaders to address the major issues that emerged from a second global war. The leaders used the paradigm to solve the routine and extraordinary problems revealed by the war.[45] The logic of their approach made particular sense because, consistent with Kuhn, the overseas paradigm provided

---

[43] See National World War II Museum, "By the Numbers: The U.S. Military," web page, undated-a; also Naval History and Heritage Command, "U.S. Navy Personnel in World War II: Service and Casualty Statistics," web page, April 28, 2015.

[44] TheUSAHEC, "World War II Basic Training," video posted to YouTube, September 10, 2009; Stecha2, "Kill or Be Killed," video posted to YouTube, August 22, 2008; Bbabbbakk, "WWII US Marines Training," video posted to YouTube, May 5, 2011.

[45] Kuhn, 2012, pp. 24, 34. In the context of Kuhn's argument, the nation's leaders were practicing *normal science*. See Chapters 3–4 for a full discussion of normal science.

more than criteria for identifying vexing post-war problems, or *anomalies*, as Kuhn would call them; more important, the paradigm itself contained the solutions to those problems.[46]

The Overseas Paradigm: Accounting for the Second World War's Anomalies

Coming out of the Second World War, the overseas paradigm needed to account for three major anomalies. First, and most significant, did the atomic bomb invalidate the prevailing assumptions about the principal location for the common defense and the heightened personal exposure to violence faced by those in uniform? Second, how should the nation reshape its military organization to account for the lessons learned from the Second World War, including assigning responsibility for nuclear weapons? And, third, what was the future of conscription and the reciprocal relationship it represented between the U.S. government and the American people?

The physical and psychological implications of nuclear weapons required leaders to address how atomic weapons—the 20th century's most profound military innovation—fit into the prevailing understanding of organized violence, since the overseas paradigm assumed proximity to violence to be a key characteristic of military service. Clearly, the atomic bomb offered a means to inflict great violence. Those serving in the military would continue to assume greater risk in the mastery and management of organized violence; however, if global nuclear war occurred, members of the military and civilians alike would suffer from an unfathomable scope of destruction. Theoretically, this was a significant definitional change, but the day-to-day responsibility for the mastery and management of violence continued to rest squarely on the military. Military members were on the front line of this new cold war in Western Europe, East Asia, and the United States. The overseas paradigm's personal connection to organized violence merely

---

[46] Kuhn, 2012, p. 37. Kuhn's formal definition for *anomaly* is "nature has somehow violated the paradigm-induced expectations that govern normal science" (p. 53).

expanded to include the existential violence of nuclear weapons that threatened physical destruction of the United States, as well as the American military.

Similarly, legislation reveals much about how the military services, as the national organizations responsible for organized violence, emerged from the Second World War. A look at the National Security Act of 1947 (NSA'47) and its 1949 amendments reveals that, after considerable debate, the services remained independent and largely autonomous. Most notably, but consistent with the overseas paradigm, NSA'47 formally created an independent air force. The Army Air Forces was already autonomous in all but name, and its performance during the war guaranteed the establishment of an organization comparable to America's Army and Navy to exercise military control, and use, of the atmosphere. Hence, the new Air Force was formally carved away from its parent service, the Army, and given responsibility for offensive and defensive air operations.[47]

Less clear, initially, was whether the responsibility for offensive air operations included assignment of organizational responsibilities for the mastery and management of nuclear weapons. This became one of the most contentious issues to emerge from the war. The newly independent Air Force, the Navy, and, to a lesser extent, the Army waged a bitter bureaucratic fight over which service would have primacy for nuclear weapons. Initially, the Air Force won this organizational struggle with the creation of the Strategic Air Command and Congress's support for an intercontinental bomber, the B-36, at the expense of a new aircraft carrier.[48] But within a decade it was clear that nuclear weapons would be central to how the Air Force and the Navy understood their responsibilities for the mastery and management of organized violence.

Additionally, two important post–Second World War developments indicated that the service autonomy of the overseas paradigm was potentially at risk: efforts to restructure the defense establishment,

[47] Public Law 80-253, National Security Act of 1947, July 26, 1947, as amended through Public Law 110-53, August 3, 2007.

[48] Rearden, 2012, pp. 82–83.

including the appointment of a Secretary of Defense and a chairman of the JCS, and the creation of the Outline Command Plan. The restructuring efforts reflected the desire to address the Second World War's sweeping command-and-control challenges. To help address them, President Roosevelt had established an ad hoc JCS and appointed Admiral William Leahy as his de facto senior military advisor during the war. While imperfect, these initiatives recognized the global nature of the conflict and signaled the need for interservice coordination and cooperation at the highest level.[49]

In a similar vein, after the war, when it came to seeking military advice, President Harry Truman made it known that he preferred to work with one civilian secretary and one uniformed advisor. Rather than consulting with separated military service chiefs, the uniformed advisor would replace the war's informal JCS advisory group. Legislative and institutional resistance from the Navy, in particular, drove Truman to accept a compromise on this, and the JCS persisted largely in its Second World War form with a new chairman of the JCS position created to serve as the JCS titular head.[50]

During the Second World War, the management of violence at the theater level also received considerable presidential attention. The importance of a single individual having responsibility, or unity of command, for the management of violence became increasingly apparent as the United States led huge offensive operations in vast theaters. To President Truman and other senior leaders, the competition between MacArthur and Nimitz for command in the Pacific theater compared unfavorably with Eisenhower's responsibility for theater command in Europe. As a result, shortly after the war, Truman supported the identification of single commanders associated with specific overseas geographic responsibilities. In addition, Truman authorized the com-

---

[49] Douglas T. Stuart, *Creating the National Security Act: A History of the Law That Transformed America*, Princeton, N.J.: Princeton University Press, 2008, pp. 52–54; Walter R. Borneman, *The Admirals: Nimitz, Halsey, Leahy, and King—The Five-Star Admirals Who Won the War at Sea*, New York: Little, Brown and Company, 2013, pp. 268–269, 436.

[50] Amy Zegart, *Flawed by Design: The Evolution of the CIA, JCS, and NSC*, Stanford, Calif.: Stanford University Press, Chapters 4–5.

mander of the newly created Strategic Air Command to assume responsibilities for strategic assets, especially nuclear weapon, without adhering to a regional affiliation. These new command arrangements were codified in the 1946 Outline Command Plan.[51] Importantly, though, even with this new command structure and the formal recognition of the JCS, legislative and executive compromises meant that the military services still retained their autonomy and their preeminent role in the monopoly, and management of organized violence held fast.

In the aftermath of the Second World War, and all of the upheavals associated with it, the reciprocal relationship epitomized by conscription also received considerable attention. General George C. Marshall, the Army Chief of Staff, in particular wanted to deliberately, but swiftly, demobilize millions of service members. Simultaneously, he wanted to create the ability to remobilize quickly and without the chaos he experienced twice in his career. Marshall saw universal military training (UMT) as the solution to the remobilization challenge.[52] UMT would require able-bodied male citizens to undergo six months of basic training and then allow them to return to the civilian world. In the event of a national emergency, they could be recalled to arms. Because of their previous training, they would already possess a level of military effectiveness unseen in previous conflicts.[53]

Although Marshall won the support of President Truman, with the end of the war and no immediate conventional threat in sight, legislative, institutional (including the Navy and Marine Corps), and popular inertia undermined efforts to pursue UMT and eventually led

---

[51] This is the forerunner to the current Unified Command Plan. Ronald H. Cole, Walter S. Poole, James F. Schnabel, Robert J. Watson, and Willard J. Webb, *The History of the Unified Command Plan, 1946–1993*, Washington, D.C.: Joint History Office, 1995, pp. 11–16. To address this overlap in mission, Truman invested considerable time and effort in passing legislation that became known as the National Security Act of 1947.

[52] Rearden, 2012, pp. 61–62; see George C. Marshall, *Memoirs of My Service, 1917–1918*, Boston: Houghton Mifflin, 1976, especially Chapter 1, for some of Marshall's personal insights into the challenges of mobilization in the First World War. The UMT concept dated back to the First World War; see Chambers, 1987, Chapter 3, in particular.

[53] Steven L. Rearden, *The Formative Years*, Washington, D.C.: U.S. Government Printing Office, 1984, pp. 13–14.

to its abandonment.[54] Unlike the end of the First World War, however, the nation passed legislation requiring all men between the ages of 18 and 26 to register for the draft and allowing the nation to call them up for active duty for up to 21 months.[55] Hence, after considerable debate, the reciprocal relationship between the U.S. government and the American people remained largely unchanged, even slightly strengthened, in the first decade after the Second World War.

Finally, the overseas paradigm needed to account for any possible impact that nuclear weapons might have on the martial organizational cultures of America's services. Would nuclear weapons require a different organizational culture to handle them? After a fitful start, the Air Force's Strategic Air Command embraced and inculcated its airmen with the importance of executing its nuclear mission on a moment's notice. When General Curtis LeMay took over Strategic Air Command, in 1948, he stressed the importance and unforgiving nature of this mission and demanded the long-standing commitment to traditional military values of mastery of violence, acceptance of hierarchy, and obedience.[56]

So even in this new nuclear age, the focus on managing violence based on unique service cultures retained a common commitment that remained inviolate. Despite distinct aspects of their organizational cultures, all the services remained committed to the martial values of operating in a hierarchy, being prepared for inflicting and suffering from organized proximate violence, and understanding that the unit's needs must come before the individual's. Perhaps the most public symbol of this common cultural commitment was the establishment of the U.S. Air Force Academy. Less than a decade after the end of the Second World War, Congress and the President established an academy for the nation's newest service. Despite the differences in conducting land

---

[54] Rearden, 2012, 39, 61–62; John Sager, "Universal Military Training and the Struggle to Define American Identity in the Cold War," *Federal History*, No. 5, January 2013.

[55] Chambers, 1987, pp. 255–257, 270; U.S. Congress, conference committee report for Selective Service Act of 1948, Washington, D.C., June 19, 1948.

[56] Warren Kozak, *LeMay: The Life and Wars of General Curtis LeMay*, Washington, D.C.: Regnery Publishing, 2009, pp. 279–314.

and air operations that led to the creation of an independent air force, the new academy's founders modeled it especially on the U.S. Military Academy, its historical antecedent. Culturally, then, the U.S. military remained largely unchanged in the first years after the Second World War. Thus, the United States and its military remained comfortable with the broad outlines of the overseas paradigm as the way to conceive of organized-violence solutions to national problems.

The Overseas Paradigm: Accounting for Additional Anomalies, 1950–1975

While geographic overlap and technological innovation hinted at future changes to the overseas paradigm, efforts to accommodate major anomalies largely succeeded through the Korean and Vietnam wars. The problems that the nation addressed in these Cold War conflicts remained consistent with the overseas paradigm—how best to create, move, and use a military overseas. The answers remained similar as well. The military services and their operational responsibilities were still defined by physical geographic spaces; the use of conscription to fill out the ranks, albeit with increasing exemptions, remained essential; and a culture devoted to organized violence and its associated personal risks remained at the services' cores. By the end of the Vietnam War, though, the nation's leaders explicitly and publicly questioned some key components of the overseas paradigm—specifically, the role of conscription and the viability of the military's traditional culture.

The overseas paradigm offered sufficient explanation for understanding the relationships among the U.S. government, the American citizenry, and their military when dealing in the realm of organized violence during the Korean War and into the early years of the Vietnam War. The Vietnam War did eventually challenge both the reciprocal relationship and the traditional military cultures when it came to organized violence. These anomalies violated the paradigm-induced expectations of the overseas paradigm. But, as Kuhn explained, a far-reaching, confident paradigm could find considerable capacity to accommodate such phenomena. It did not look for novelties of fact and theory, and, at its most successful, uncovered none. Among other things, the power of the paradigm drove the construction of elaborate equipment and the development of a specialized vocabulary and skills. Even when something seemed to go wrong, the phenomenon could be explained within the context of the existing paradigm.[57]

Conscription helped populate the military's ranks from the end of the Second World War through Vietnam War. However, the popular distrust of the government in general and the military in particular spawned by Vietnam ultimately made it impossible to sustain the draft. As Beth Bailey documents in *America's Army*, several factors shaped this decision, including debates over the responsibility to serve, the economics of service, the need for better racial and gender integration, and the organizational dysfunction of America's military. After myriad debates and studies, as well as the passage of necessary legislation, Secretary of Defense Melvin Laird formerly ended the draft in January 1973. Thus, even though young men continued to register as part of the Selective Service System, the idea of national mobilization disappeared, replaced by the notion that volunteering was an essential precondition for military service.[58]

One of the main catalysts for this shift to an all-volunteer force was the breakdown of the military's organizational culture during the Vietnam War. The widespread antipathy against the military generated by the Vietnam War found its way into the military. Lieutenant General Karl Eikenberry (Ret.) recounted that, as a West Point cadet in the mid-1970s, he entered some overseas barracks as part of his officer training only to discover a combat-ineffective world, one rife with racial tensions and drug abuse.[59] The profound breakdown in discipline he witnessed reflected a larger rejection of the traditional

---

[57] Kuhn, 2012, pp. 52–53, 64–65, 76.

[58] Beth Bailey, *America's Army: Making the All-Volunteer Force*, Cambridge, Mass.: Harvard University Press, 2009, Chapter 1; see also Bernard D. Rostker, *I Want You! The Evolution of the All-Volunteer Force*, Santa Monica, Calif.: RAND Corporation, MG-965-RC, 2006, pp. 1–9; David R. Segal, *Recruiting for Uncle Sam: Citizenship and Military Manpower Policy*, Lawrence: University of Kansas Press, 1989.

[59] James Kitfield, "The Great Draft Dodge," *National Journal*, December 13, 2014.

martial culture by many, but certainly not all, Army units. Not surprisingly, for individuals in units similar to the one Eikenberry visited, discipline, hierarchy, sacrifice, and other traditional values similarly dissipated as core organizational qualities.

As the Vietnam War came to a close, the overseas paradigm had accommodated some significant anomalies. The relationship between the U.S. government and the American people had fundamentally changed as a result of Vietnam conflict. For the first time in decades, the United States chose to rely on an all-volunteer force. The paradigm, however, easily accommodated this change. It only needed to look back to the years between the two world wars. During those years, the United States had quickly abandoned the draft because of the lack of national threat. Similarly, with the exception of long-standing commitments to South Korea and NATO, in Europe, there was little appetite for overseas involvement in the years after Vietnam. So eliminating the draft, at least in one sense, was a return to an older model for military service. The move to an all-volunteer force also provided a means to restore traditional military values. By self-selecting to put on a uniform, those agreeing to military service were incentivized to adhere to the traditional organizational values that many Vietnam-era draftees rejected. As America moved into a post-Vietnam era, despite all the tumult, the overseas paradigm managed to accommodate, or at least rationalize, the anomalies and remain in place.

The Overseas Paradigm: Conflicting Signals, 1975–2001

After the Vietnam War ended, the focus of America's military remained overseas. The concept of organized violence remained unchanged even as the Department of Defense started to invest time and capital into new technologies to offset the vast size of the Soviet military. These technologies focused on increasing the stealth, range, and precision of America's military weapons to give U.S. forces stationed in Europe a qualitative advantage until they could be reinforced. The personal expectations of experiencing organized violence were still central to how American leadership understood the relationship between organized violence and the nation-state.

Some major anomalies did challenge the overseas paradigm in the years between Vietnam and the 9/11 attacks. These anomalies resulted from confused, even failed, military operations; initiatives to adopt the corporate practices of outsourcing and privatization; and efforts to mirror the changes in American society. First, in the decade after Vietnam, a series of bungled contingency operations involving Iran, Lebanon, and Grenada ultimately drove the most-sweeping organizational changes since 1947. Among other issues, Congress found in the failed rescue attempt of American hostages in Iran (known in the military as Desert One) profound problems with interservice coordination. Specifically, post-operation assessments pointed to service independence and the lack of proficiency in joint operations as the causes of failure. Even the so-called successful operation in Grenada a few years later reinforced the extent of these problems.[60]

Frustrated with the military's inability to reform from within, Congress subsequently took the lead in drafting legislation that would end service autonomy in military operations. This legislation, titled the Goldwater-Nichols Act, made the combatant commanders responsible for military operations. It did so by putting them in charge of matrix organizations relying on the resources and manpower from two or more services. The legislation ensured the combatant commanders' authority by directly placing them in the reporting chain of command to the Secretary of Defense and the President. Furthermore, the legislation clearly stated that, although not in the chain of command, the chairman of the JCS served as the President's principal military advisor, not the entire JCS. The services, in short, lost their direct access to the President and Secretary of Defense, as well as their operational autonomy, when this legislation passed in 1986. Based on the lessons from Desert One, Congress also used the Goldwater-Nichols legisla-

---

[60] James R. Locher III, *Victory on the Potomac: The Goldwater-Nichols Act Unifies the Pentagon*, College Station: Texas A&M University Press, 2004, pp. 45–48, 127–132, 135–136, 141–163, 305–313.

tion to mandate the creation of SOCOM, with its own four-star general and budget authority.[61]

Second, within a few years of the Goldwater-Nichols Act, the end of the Cold War encouraged the dramatic reduction in the size of the military overseas and in the United States, leading to pressures to outsource and privatize military functions to save money. This initiative jolted the military. Among other things, this anomaly was documented in the recommendations made by the Commission on Roles and Missions (CORM) in the mid-1990s. CORM was a congressional response to two different imperatives—reaping the benefits of the end of the Cold War and bringing best business practices to the U.S. military. It was in this latter category that CORM highlighted the changes in the overseas paradigm. Many of the commission's recommendations focused on improving business practices, especially by outsourcing those activities that were not inherently military.[62] The CORM report simply reflected the conventional wisdom of the day that held that, wherever possible, military support activities should be outsourced. The commission noted:

> More than a quarter of a million DOD employees engage in commercial-type activities that could be performed by competitively selected private companies. Experience suggests achievable cost reductions of about 20 percent. DOD should outsource essentially all wholesale-level warehousing and distribution, wholesale-level weapon system depot maintenance, property control and disposal, and incurred-cost auditing of DOD contracts. In addition, many other commerical-type

[61] United States Special Operations Command, *History of United States Special Operations Command*, 6th ed., MacDill Air Force Base, Fla., March 31, 2008, pp. 6–7.

[62] John P. White, Antonia H. Chayes, Leon A. Edney, John L. Matthews, Robert J. Murray, Franklin D. Raines, Robert W. RisCassi, Jeffrey H. Smith, Bernard E. Trainor, and Larry D. Welch, *Directions for Defense: Report of the Commission on Roles and Missions of the Armed Forces*, Washington, D.C., U.S. Government Printing Office, 1995, recommended two major opportunities be pursued aggressively: implementing the long-standing national policy of relying primarily on the private sector for services that need not be performed by the government and reengineering the remaining government support organizations. See especially Chapter 3.

> activities, including those in family housing, base and facility maintenance, data processing, and others could be transferred to the private sector. Finally, DOD should rely on the private sector for all *new* support activities.[63]

Eventually, the government-wide emphasis on military downsizing, outsourcing, and privatization resulted in contractors replacing military members in so many activities that the latter could not do their mission without contractor support. Most notably, by the mid-1990s, lacking sufficient manpower, the U.S. military could not support training and security activities in the Balkans, relying instead on a privatized military contractor called Military Professional Resources Incorporated (MPRI).[64]

Finally, the military of this era had to deal with the significant social change demanded by race, gender, and gay-rights issues. Since all service members were now volunteers, the services could inculcate new members with traditional military values in a way that many of their Vietnam-era draftee predecessors rejected. However, pressured by Congress to be more inclusive, and worried about sufficient manpower to fill the ranks, the military services realized that they had to pay closer attention to social issues. Initially, much of the effort focused on ensuring successful racial and gender integration. The latter in particular sparked controversy because of a perceived decline in military effectiveness as the military expanded the number of fields open to women, even while they were excluded from combat. A related controversy ensued early in the Bill Clinton era, as the new President pressed to allow openly gay and lesbian individuals to serve in the military. When the uniformed leaders once again decried the impact of this social issue on military effectiveness, the President backed down and accepted a compromise policy known as "don't ask, don't tell."[65]

---

[63] White et al., 1995, p. ES-6; emphasis in the original.

[64] E. B. Smith, "The New Condottieri and US Policy: The Privatization of Conflict and Its Implications," *Parameters*, Winter 2002–2003, p. 110.

[65] National Defense Research Institute, *Sexual Orientation and U.S. Military Personnel Policy: An Update of RAND's 1993 Study*, Santa Monica, Calif.: RAND Corporation, MG-1056-OSD, 2010, pp. 39–47

Despite these many assaults on the overseas paradigm, its underpinning tenets—that a national military, divided into services responsible for physical geographic spaces and committed to operating in physically violent, demanding environments—continued to shape how Americans envisioned military solutions to national problems at the end of the 20th century. The paradigm was under siege on all fronts, but—as noted, and consistent with Kuhn—with additional explanations and qualifications, the existing paradigm still held. The post-9/11 world would shatter the overseas paradigm.

Inconsistencies in the Overseas Paradigm: The Post-9/11 Era

With the 9/11 attacks, the overseas paradigm faced additional anomalies that were even more difficult to explain and incorporate. The increased scale of, and access to, means of violence by nonstate actors allowed transnational insurgent groups access to formerly inaccessible means of violence. As the United States tried to cope with violent extremists, it relied heavily on the military. By 9/11, the national organizations responsible for organized violence had changed significantly. The combatant commander matrix had solidified, putting CENTCOM in charge of ongoing military operations and prompting General Franks's Title X remark when the JCS questioned the viability of his operations plan. When the tasks at hand became too much for CENTCOM's commander, subordinate commands in Iraq and Afghanistan were created, arguably moving the services even more to the periphery of current military operations. The CIA's paramilitary capabilities placed it at the forefront of early military operations in Afghanistan. Clearly, the traditional autonomy of the military services had been supplanted by the combatant commander matrix organizations, and, at least early on, the services were deemed largely irrelevant compared with these post-9/11 specialized military and paramilitary operations.[66]

---

[66] Bradley Graham, *By His Own Rules: The Ambitions, Successes, and Ultimate Failures of Donald Rumsfeld*, New York: Public Affairs, 2009, pp. 300–315.

From the start of the global war on terror, it was also evident that incentivized relationships continued to replace reciprocal ones between the U.S. government and the American people. Especially as the operations in Iraq and Afghanistan increased in tempo, rather than look to a draft to expand the force, uniformed military volunteers received multiple incentives to remain in uniform. These included an expanded G.I. Bill, military pay raises that exceeded civilian counterparts, and additional compensation for those serving in a theater of operations.

The incentivized relationship carried over into the civilian realm, where adherence to civilianization, outsourcing, and privatization over the past 20 years produced a large, diverse government civilian and civilian-contractor force. Some responsibilities for strategy, planning, and logistics that once belonged to the uniformed military shifted to government civilians. Even more dramatically, employees at such companies as Kellogg, Brown and Root, CACI, and Blackwater frequently accepted the personal dangers associated with organized violence as a term of employment, whether performing base support functions or providing personal security. Paradoxically, this occurred while critical military responsibilities either shifted back to or became rooted in U.S. territory, where military personnel operated from the safety of U.S. bases. Moreover, while the press kept a tally of every military casualty, contractor casualties drew little attention from the U.S. government or the media. When civilian contractors died in a war zone, it was easier for the nation to look the other way. The government did not have to respond, and popular reaction, if any, tended toward callousness, along the lines of: They knew it was dangerous when they took the job and the large paycheck.[67]

[67] Peter W. Singer, "Can't Win with 'Em, Can't Go to War Without 'Em: Private Military Contractors," Brookings Foreign Policy Paper Series, Washington, D.C.: Brookings Institution, September 2007; Peter W. Singer, "Counterproductive," *Armed Forces Journal*, November 1, 2007. The Congressional Research Service reported that, from 2007 to 2011, contractors provided more than 50 percent of the Department of Defense's workforce (defined as uniformed personnel and contractors) in Afghanistan and 50 percent in Iraq. See Moshe Schwartz and Joyprada Swain, *Department of Defense Contractors in Iraq and Afghanistan: Background and Analysis*, Washington, D.C.: Congressional Research Service, May 13, 2011, p. 2.

Moreover, military culture itself changed with emergence of non-traditional and, until recently, technically impossible activities relative to the overseas paradigm, including remote, long-term surveillance; remote killing; and cyber destruction. New technology and business practices gradually distanced many in America's military from a close proximity to organized violence, so the immediate relevance of hierarchy, discipline, fitness, and courage became less evident. Satellite, drone, and cyber technologies, for example, allowed the U.S. government to acquire vast amounts of information about possible adversaries and disrupt their operations from American soil. Drone operations epitomized this changing dynamic. With the advent of a large, capable drone force, military members could conduct assigned surveillance and lethal operations at no personal risk, operating in relative physical comfort from locations that were even safer than civilian neighborhoods.[68]

The discussions of cyber war further highlighted this conundrum. Taken to a theoretical extreme, at least in the public's imagination, cyber experts could remotely flip switches or unleash computer codes, thereby anonymously killing scores to thousands of people, including the young and elderly. These cyber experts might be military, but they could just as well be civilian, and they might be subject to organized-violence reprisal attacks (but most likely not). In any event, even more than the drone operators, in theory, they would also be delivering organized physical violence on behalf of the nation in safety and anonymity.[69] Any act of retaliation, in fact, would probably target, and almost certainly affect, others rather than the specific cyber experts. Hence, to some in the military, the martial qualities do not apply to this group, and they hypothesize that these qualities inhibit, and even undermine,

---

[68] For detailed discussions of drone operations and their implications, see, for example, Gregoire Chamayou, *A Theory of the Drone*, New York: The New Press, 2013; Andrew Cockburn, *Kill Chain: Drones and the Rise of High-Tech Assassins*, New York: Henry Holt and Co., 2015. There is also a growing academic and popular literature specifically on ethical aspects of remote warfare; see, e.g., Bradley Strawser, *Killing by Remote Control: The Ethics of an Unmanned Military*, Oxford, UK: Oxford University Press, 2013. Also see Daniel Brunstetter and Megan Braun, "The Implications of Drones on the Just War Tradition," *Ethics and International Affairs*, Vol. 25, No. 3, 2011.

[69] John Hamre, "The Electronic Pearl Harbor," *Politico*, December 9, 2015.

the creative culture essential to successful cyber operations. Collectively, these impressions contribute to a common stereotype in some military circles that the best cyber operators are younger, entrepreneurial civilians from the millennial generation, who would wear jeans and hoodies to work.[70]

Finally, even as the traditional military culture shifted, social pressures to accommodate larger cultural shifts in American society increased. Senior military leaders devoted considerable time, amid two wars, to adjusting to these shifts, which included repealing "don't ask, don't tell"; reevaluating combat exclusion for women; and potentially issuing a new transgender policy. This caused some military personnel to wonder, sardonically, whether the military's purpose was to master and manage organized violence or pursue social change.[71]

Crisis and the Overseas Paradigm

The inconsistencies in the overseas paradigm almost 15 years after the 9/11 attacks show how much the American military and its relationship to the government and the American population have evolved since this paradigm emerged out of the First World War. Once, the military understood where it operated, how to think about organized violence, and the reciprocal relationship to the American people (as revealed during the First World War), but now everything is much more complex, confused, and confusing.

[70] This is not to imply the military lacks talented cyber operators, indeed they are part of the cyber mission force created by the Department of Defense, in December 2012. Rather, this observation reflects multiple conversations with serving and retired officers regarding what types of individuals seem to be best suited to conduct cyber operations over the long term. The cyber mission force, once fully operational, will consist of 6,200 military, civilian, and contractor personnel. See U.S. Department of Defense, *The DoD Cyber Strategy*, Washington, D.C., April 2015. It is also worth noting that, in October 2015, the General Services Administration released a five-year, $460 million multiple-award request for proposals (RFP) to outsource Cyber Command's mission support; see Cheryl Pellerin, "Rogers: Data Manipulation, Non-State Actor Intrusions Are Coming Cyber Threats," U.S. Department of Defense, November 19, 2015.

[71] This point came up frequently in informal conversations with serving military officers.

Location. America's understanding of the core location of the common defense is in flux. Although the overseas physical geographies of the Western Pacific, the Middle East, and Europe remain important, they only partially cover America's emerging concept of the common defense. Policymakers expanded the common defense to include American lives and livelihoods in the United States that can be held at risk in a world of global terrorism, as well as in the boundaryless worlds of space and cyberspace. The dominant locations for the common defense simultaneously exist overseas, on the North American continent, and in cyberspace's amorphous world. Conceptually, today's common defense lacks physical geographic boundaries—it is ubiquitous.

Concept of violence. The U.S. government, the American people, and their military no longer have a shared understanding about the nature of organized violence, because of the rise of nonproximate violence. Organized violence has changed so much in the past 15 years that personal, proximate risk is only one aspect of how Americans think about using organized violence on behalf of the nation. Indeed, if media reporting is an indicator, in the realm of nonproximate violence, there are at least three variations, or dimensions, to consider. The first variation is the partner-nation proxy violence encouraged by the COCOMs. While the idea of training and equipping partner nations is not new, the extent to which it is being pursued as central to the common defense is new. Most notably, in addition to the regional COCOMs, as an article in the *Wall Street Journal* recently observed, "These days, the sun never sets on America's special operations forces. Over the past year, they have landed in 81 countries, most of them training local commandos to fight so American troops don't have to."[72] The implicit assumption throughout this effort is that improving partners' ability to fight allows the United States to increase its own warfighting capacity. Thus, a large emphasis on building partner-nation militaries becomes a de facto means to rationalize a smaller U.S. military.

---

[72] Michael E. Phillips, "U.S. Commandos Span the Globe," *Wall Street Journal*, April 25–26, 2015.

A second dimension of nonproximate violence receiving significant media attention is the remote violence of drone warfare and the confused nature of its conduct. President Barack Obama's administration, for example, revealed in late April 2015 that two Western hostages were killed by U.S. drone strikes targeting al Qaeda militants in Pakistan. Drone strikes such as these highlight the intertwined relationship between the CIA and the military when employing remote violence. This specific example points to the continued American domestic discomfiture associated with remote violence and the dronization of warfare.[73] It also raises uncomfortable questions about how the United States should think about individuals who inflict organized violence on others from a remote, protected location. For example, are these remote American drone operators—flying their unmanned aerial vehicles from air-conditioned trailers thousands of miles from any conflict—pilots, warriors, intelligence experts, assassins, cowards, or all of these? Should they even serve in the military? Or do they belong in the CIA, or in some new, yet-to-be-created organization?

A third variation of nonproximate violence is the theorized anonymous, even devastating, violence of cyber warfare. In defense policy circles, at least, cyber operations are conceptualized as something that can destroy everything from defenseless American citizens—by contaminating their water supply, for example—to the nation's economy. This compels senior policy leaders, such as Leon Panetta, to invoke the possibility of a "cyber Pearl Harbor" to drive home the importance of cyber operations.[74] In part because of the visceral power of this conceptualization, Secretary of Defense Ashton Carter's cybersecurity strategy resulted in substantial media attention, including an editorial in the *New York Times* that credited the strategy with bringing transparency to a "military program that is expected to increase to 6,200 workers in a few years and costs billions of dollars annually." The editorial went on to say: "It is essential that the laws of armed conflict

---

[73] Adam Entous, "Obama Kept Looser Rules for Drones in Pakistan," *Wall Street Journal*, April 27, 2015; Chamayou, *A Theory of the Drone*, pp. 185–194.

[74] Leon Panetta, *Worthy Fights: A Memoir of Leadership in War and Peace*, New York: Penguin Press, 2014, pp. 433–434. See also Hamre, 2015.

. . . are followed in any offensive cyberoperations."[75] The cybersecurity strategy assumes that nonproximate violence cyber operations, as well as defensive and information technology–oriented efforts, are now central to the military's purpose, even though a substantial number of its practitioners are civilians. Is this logical?

In short, the partner-nation proxy violence encouraged by the COCOMs, the remote violence of drone warfare, and the theorized anonymous violence of cyber war garner as much attention as the traditional military responsibilities for personal, proximate organized violence. What is both disconcerting and intriguing to consider in this realm of nonproximate violence is the looming question, "What's next?" Will it be killer robots or some other form of nonproximate warfare that will add to this conundrum of how to understand organized violence, and who or what organizations will be responsible for it?[76]

These nonproximate ways that the American military employs different types of violence suggest that organized violence is entering a new era. If this trend continues, the American military will have embraced a more encompassing concept of organized violence, including proxy, remote, and anonymous violence. The fundamental precondition that has underpinned the American military since its inception, that inflicting violence on behalf of the nation requires greater personal risk, no longer holds to the same extent for some, and perhaps most, who wear a military uniform. Thus, in the past 15 years, the military has passed some undefined but vital threshold for inflicting and experiencing organized violence that fundamentally challenges its core purpose.

Organization and people. The overseas paradigm's increasing inability to accommodate anomalies is reflected in corresponding confusion over how best to organize the military. Although physical geography made sense in the early decades of the 20th century, the scope of conflict and technological improvements suggests that other organiza-

---

[75] This is a reference to the cyber mission force. Editorial Board, "Preparing for Warfare in Cyberspace," *New York Times*, April 28, 2015; U.S. Department of Defense, 2015.

[76] Brian Fung, "Elon Musk and Stephen Hawking Think We Should Ban Killer Robots," *Washington Post*, July 28, 2015.

tional options based on functional expertise might be in the offing for the 21st century. This explains de facto permanent, military service–equivalent organizations, such as SOCOM and Cyber Command. The emergence of remote, anonymous violence further complicates the military's chains of command. Most notably, the Afghanistan and Iraq examples highlight that geographic combatant commanders are at least as focused on partnership building and proxy operations by partner nations as they are on combat operations. Perhaps geographic combatant commanders themselves are becoming anachronistic. Efforts to combat the Islamic State further highlight these complicated relationships. No fewer than three combatant commanders have crucial responsibilities for this fight, not to mention the task force commanders and other officials involved in this mission.[77] This leads to complex chains of command that make it difficult to assign accountability for the success or failure of a military operation.

Where the manpower for these operations will come from is similarly unclear. Not only is the draft no more but even those advocating its restoration do so to cultivate a commitment to national service rather than to fill the military's ranks.[78] Absent a calamity, the all-volunteer force will remain, but with insufficient manpower to perform its task. In the 21st-century American military, it is much easier to contract than to conscript. That is why using well-paid contractors, even in a war zone, is acceptable; initiating a draft is not. As the military services seek more expertise in nonproximate violence in the drone and cyber realms, this reliance on civilian contractors will likely remain and even grow.[79]

Organizational culture. The military's organizational culture has fractured and splintered dramatically. This situation is closely tied to

---

[77] Take, for example, the ongoing operations against the Islamic State that involve CENTCOM, European Command, Africa Command, SOCOM, U.S. Army Central (as an operational commander), and the President's Special Envoy.

[78] Stan McChrystal, "Beyond the Draft: Rethinking National Service," *Defense One*, November 29, 2015.

[79] See, for example, W. J. Hennigan, "Air Force Hires Civilian Drone Pilots for Combat Patrols; Critics Question Legality," *Los Angeles Times*, November 27, 2015.

the changing understanding of violence. Once, organized violence and its associated shared personal risk determined the essence of military culture; now, it is difficult to discern a single cultural identifier for America's military. What might be suited for those in special operations carries little resonance for those in cyber or space operations, much less the acquisition corps. Even the symbolism of the military as a physically isolated community continues to erode as the services look for additional ways to divest themselves of base infrastructure. Military bases used to be self-contained communities; now many are looking to surrounding civilian communities, large and small, to provide housing, utilities, and community services. Indeed, every cultural aspect of the overseas paradigm has been breached—physical isolation; living and operating in austere, often dangerous, environments; the necessity for physical stamina and courage; and, most fundamentally, the responsibility to master some aspect of violent means that inherently subject the individual to more-hazardous circumstances.

Incommensurability. Thus, in ways that would be unrecognizable to those who served in the First World War, organized violence now tends toward the nonproximate, even abstract, as much as the personal and proximate. The military's fundamental purpose to master and manage organized proximate violence on behalf of the nation has dissipated, along with the overseas paradigm. In Kuhn's terms, the U.S. military today faces an incommensurable problem.[80] Although the concepts, organizations, and language of the overseas paradigm remain in use, they no longer sufficiently explain how the U.S. government understands its military, how the American people relate to it, or even what the military's purpose is in 21st-century conflict. Comparable to the Ptolemaic paradigm, the overseas paradigm fails without complicated, convoluted rationalizations and explanations. Kuhn would argue that this is a huge, profound, and unrecognized issue.[81]

---

[80] Kuhn, 2012, p. 147.

[81] Kuhn, 2012, pp. 90, 92, 103, 129, 135–142, 148.

Rise of the Guardian Paradigm?

The overseas paradigm no longer explains how the United States understands the common defense. Every aspect of the paradigm has failed, and even the lexicon used to discuss the military's role in the common defense is inadequate. Kuhn argued that paradigms are in crisis when they can no longer easily accommodate anomalies.[82] If the anomalies identified above indicate a crisis in the overseas paradigm, then it is time for the United States and its military to address some first-order questions: What is the location of the common defense? What constitutes organized violence in the 21st century? How much of it should be the responsibility of the military? What elements belong elsewhere? What does it mean to serve in the military—who is included and who is excluded?

The answers to these questions shape what type of organizations and individuals the nation needs to provide for the common defense. They delineate who is in the military and who is doing something else important for the nation, such as serving in new types of public institutions, but not ones focused on organized violence. Right now, the nation avoids these choices. It uses existing military organizations, budgets, and people to take on a variety of tasks. Through bureaucratic shuffling, redrawing lines of organizational command, control, and responsibility, the military adapts to new tasks as they emerge. Sometimes adaptation amounts to making a logical choice; other times it means doing what is expedient. When it is the latter, what the military does is diluted by making it the keeper of additional capabilities that arguably should be done elsewhere. Paradoxically, this approach, while diluting the military's core skill sets, gives the military authorities and responsibilities that perhaps it should not have. Cyber operations emerge as the most notable candidate for discussion. Why is the NSA in the Department of Defense, much less dual-hatted as Cyber Command, when much of what the NSA does is domestically focused? Even if this made sense before cyber technology became so pervasive—is it still the right answer more than 60 years after the NSA's creation? A

---

[82] Kuhn, 2012, Chapter 8, especially p. 82.

similar set of questions could be asked about the operation and sustainment of America's space constellations and several defense agencies.[83]

So today's U.S. military is caught between two irreconcilable positions. On the one hand, it uses the overseas paradigm's language and values to describe its roles, missions, and responsibilities. On the other hand, the overseas paradigm is woefully inadequate to explain all that the military is now asked to do, much less the number and type of actors involved in its myriad tasks. This disconnect helps explain some of the distinct cultures that have emerged. For example, in a few cases, entire service organizations, such as the Marine Corps and SOCOM, have emphasized the importance of maintaining a martial culture and operating effectively in close proximity to organized violence. In another case, civilian contractors have created paramilitary organizations that answer to boards of directors rather than military officers. Military officers once commanded organizations with similar responsibilities; now they are limited to providing contract oversight. In a third case, uniformed, specialized technocrats have created their own organizational subcultures. These are based on unique technical responsibilities, including providing important communications-navigation information, using remote means to look for suspicious activity, acquiring military hardware, and even thinking creatively about new technology to use against adversaries. These technocrats perform vital tasks but have little connection either to organized violence or to their parent military organizations. Indeed, some of their organizations, especially in the realm of nonproximate violence, would perhaps thrive better with a distinctly nonmilitary culture to accomplish their missions. Collectively, the individuals and their organizations in the latter case could represent the vanguard of the new guardian forces—crucial to the nation but fundamentally different from their military antecedents.

The emergence of guardian forces raises two profound sets of policy issues. The first concerns the location of the common defense.

---

[83] These defense agencies could include the Defense Commissary Agency, Defense Intelligence Agency, Defense Security Cooperation Agency, and National Geospatial-Intelligence Agency, just to name a few.

Does it have a dominant location any longer, or is it boundless, ubiquitous? If the former, what are the boundaries? If the latter, by defining it so broadly, does it lose meaning? The second set focuses on the relationship between the person in uniform and organized violence. For those who provide for the common defense, does a personal connection to organized violence still matter? If a proximity to violence does not matter, why not? If it does matter, how should we think about individuals personally removed from violence but still somehow associated with it?

An emerging guardian paradigm—unheralded, uncomfortable, yet compelling—posits the following answers to these questions. The guardian paradigm (1) identifies the location of the common defense in boundless terms; (2) sees organized violence as something that can be inflicted remotely, anonymously, even robotically by individuals operating in safe, nondemanding physical environments, as well as through serving in proximity to violence; (3) depends organizationally on military, civilian, and contractor personnel and capabilities; and (4) requires a diversity of organizational cultures, rather than one built on traditional military values.

Some cyber, space, and drone operators, as well as many individuals associated with acquisition, have already become the vanguard of these new, selectively violent guardian forces. The call for more defense entrepreneurs and the desire to leverage millennial-generation talent to advance cyber and remote technologies further challenge traditional military organizations and cultures, and suggest that a paradigm shift is already under way. Ultimately, the guardian paradigm highlights the transformation of the U.S. military into a holding-company equivalent of several national security and public service capabilities, which in many cases are only loosely tied to proximity to organized violence.

If this analysis is correct, guardian forces then need to be identified, understood, resourced, organized, and developed in ways that are different from the military. Without this delineation, organizational confusion over the roles and responsibilities of America's military will persist until an existential crisis demands their explicit recognition. The nation can little afford to let a challenge of this magnitude linger. America's 21st-century leaders face a challenge the framers would have appreciated: In a world where the location of the common defense is ubiquitous, how best should the nation provide for it, and what types of organizations, military or otherwise, are necessary to accomplish this? Addressing this challenge will ultimately reset the fundamental relationship among the U.S. government, the American people, and the military's role in organized violence.

Possible Implications of a Paradigm Change

The rise of the guardian paradigm does not mean that the overseas paradigm has ceased to exist or even ceased to be the dominant way the nation still thinks about the common defense. Indeed, the nation remains in the midst of that change. Determined adherence to the overseas paradigm does mean, however, that the nation will continue to conflate organized violence and guardian responsibilities. It also will continue to explain institutional issues and look for their solutions in increasingly complex, even convoluted, terms.

Conversely, accepting the emergence of the guardian paradigm leads to some uncomfortable questions. For example, from a personnel perspective, for guardian skills, such as those in cyber, seemingly undermanned and reliant on the civilian sectors, what type of recruitment, organization, and organizational culture encourages the best in the nation to join and serve in a guardian role?[84] From a training perspective, how does initial training for the guardian forces differ from basic military training? From a budgetary perspective, how will the nation respond when it sees how much of the defense budget is actually devoted to guardian responsibilities? Will it want to spend more on defense, shift the dollars to guardian responsibilities, or perhaps increase both? From a defense reorganization perspective, should the next Goldwater-Nichols Act look at divesting responsibilities from the military rather than pursuing more interagency integration, espe-

[84] As the head of Cyber Command and NSA, ADM Michael S. Rogers noted, in November 2015, regarding the issue of private-sector help: "We turn to the private sector to harness the abilities and their capabilities to generate the tools DoD needs to execute its broad mission to defend the nation and protect our interests." See Pellerin, 2015.

cially since so much of the interagency is already resident in the military? From a warfighting perspective, what are the core organizations of organized violence? Are the COCOMs still useful? Did they provide an interim means to integrate forces operating in different geographic spaces but now more resemble the services as they advocate for resources and responsibilities? And, from the profession of arms perspective, how does this concept apply, if at all, in the guardian paradigm? Is it an archaic concept or one that remains relevant only to those personally subjected to organized violence? These are just a few of a long list of questions that the guardian paradigm raises. Once the paradigm shift is accepted, the myriad issues and questions facing the nation crystallize.

Thus, a deep dissonance exists between the overseas and guardian paradigms. Identifying what is core to the U.S. military of the guardian paradigm, and what best belongs elsewhere in the U.S. government, is one of the great policy challenges of the first half of the 21st century. In the century since the First World War, the linkages among the U.S. government, its citizens, and the American military have become complicated and confused. This is especially apparent in the American understanding of, and the military's role in, managing organized violence. The military has gradually turned into something very different. It was a collection of armed services distinguished by its mastery, management, and employment of organized violence. It provided for the common defense, first in North America and then overseas. Now many aspects of the 21st-century military are less clearly connected to the personal risk associated with organized violence, but they also suggest broader day-to-day guardian responsibilities. Unless understood and carefully managed, the scope of this change could foster organizational confusion across much of the military and confound Americans about how to think about the military and its role in the nation's defense.

Kuhn noted that the transfer of allegiance from one paradigm to another "is a conversion experience that cannot be forced." Moreover, he understood that resistance to a new paradigm "is inevitable and legitimate."[85] Paradigm shifts eventually occur because the existing paradigm can no longer accommodate major anomalies and crisis ensues. Over the past decade or so, as the United States dealt with wars in Afghanistan and Iraq, it adjusted its martial expectations to not losing wars, rather than winning them. This meant that the overseas paradigm's crisis approached more subtly, without the national trauma caused by a major military defeat, such as what France experienced in 1940.[86] But the stakes are no less real for the United States in the 21st century than they were for France 75 years ago.

The overseas paradigm persisted until the United States defined the location of the common defense in boundless terms and ceased to consider proximity to organized violence as a precondition for the mastering and managing of violence on the nation's behalf. With these changes, the overseas paradigm lost its coherence and hence its relevance. Failure to recognize this loss and adapt to the rise of guardian forces has immeasurable consequences for America's military. Civilian and military leaders need to recognize, understand, and assess what is happening to this enormously important national institution. Only then can they shape the military in a way that best serves the nation's 21st-century needs. Absent their active involvement, Americans could well look back in a few decades and discover that they are supporting an expensive national organization, military in name only, that they do not recognize, understand, or, most important, trust.

Epilogue: The Colonel's Retirement Speech

Dad, Granddad, you are my inspiration. I originally planned to highlight our many connections as brothers-in-arms, spanning multiple generations and wars. Tragically, I cannot. Through the vagaries of today's

---

[85] Kuhn, 2012, pp. 150–151.

[86] Marc Bloch, *Strange Defeat*, New York: W.W. Norton and Co., 1999, is an excellent example of a French soldier-scholar trying to understand France's catastrophic defeat by Germany in 1940.

world, you and my sister, your daughter and granddaughter, became more like brothers- and sisters-in-arms instead. This cherished status sadly came at the cost of her life.

So rather than a sweeping speech, I offer a simple confession: I long for the same clarity of understanding about my service that you possess. There is so much about today's military, my military, that doesn't make sense: how it's organized, its tenuous relationship to organized violence, what the nation really wants it to do. As I prepare to hear my retirement orders read aloud, I conclude that I served in a military in many ways as confused as I am. It is in crisis about its purpose and its expectations for those who serve.

I'm comforted by the realization that America creates and re-creates the military it thinks it needs "to provide for the Common Defense." My military reflected our nation's struggle with this seemingly innocuous, deceptively complex concept. Hopefully my service in some very small way helped the nation bridge from the military it needed in the past to the military it needs in the future to ensure our common defense. I just trust that it can build this new military, whatever it looks like, in time.

As I leave my uniform behind, I look back with gratitude for my years of service. They weren't exciting, much less dangerous. Unlike you, my "war stories" focused on fighting the bureaucracy, not the enemy. But I know these things for certain—I chose to serve, took an oath, wore a uniform, and anticipated experiencing the risks you experienced in uniform. Twenty-some years ago, I wouldn't have imagined describing my service as a small pillar in the bridge to a new military—but that's my legacy. It's unsettling in many ways, but, surprisingly, it's enough.

Publish the orders . . .

Abbreviations

AEF	American Expeditionary Forces
CENTCOM	U.S. Central Command
CIA	Central Intelligence Agency
COCOM	combatant command
CORM	Commission on Roles and Missions
IED	improvised explosive device
JCS	Joint Chiefs of Staff
NATO	North Atlantic Treaty Organization
NSA	National Security Agency
NSA'47	National Security Act of 1947
ROTC	Reserve Officers' Training Corps
SOCOM	U.S. Special Operations Command
UMT	universal military training

Bibliography

Army VideoTube, "U.S. Army Hand to Hand Combat Training," video posted to YouTube, April 9, 2013. As of December 3, 2015:
https://www.youtube.com/watch?v=wfGC8EOWABI

Bailey, Beth, *America's Army: Making the All-Volunteer Force*, Cambridge, Mass.: Harvard University Press, 2009.

Bbabbbakk, "WWII US Marines Training," video posted to YouTube, May 5, 2011. As of February 9, 2016:
https://www.youtube.com/watch?v=ojBNzdNhHrw

Bird, Alexander, "Thomas Kuhn," revised August 11, 2011, in *The Stanford Encyclopedia of Philosophy*, Stanford, Calif.: Metaphysics Research Lab, Center for the Study of Language and Information, Stanford University, 2013. As of November 30, 2015:
http://plato.stanford.edu/archives/fall2013/entries/thomas-kuhn/

Bloch, Marc, *Strange Defeat*, New York: W.W. Norton and Co., 1999.

Bobbitt, Philip, *The Shield of Achilles*, New York: Anchor Books, 2002.

Borneman, Walter R., *The Admirals: Nimitz, Halsey, Leahy, and King—The Five-Star Admirals Who Won the War at Sea*, New York: Little, Brown and Company, 2013.

Brutus, "Certain Powers Necessary for the Common Defense, Can and Should Be Limited," Anti-Federalist Paper No. 23, The Federalist Papers Project, 1788a. As of February 8, 2016:
http://www.thefederalistpapers.org/antifederalist-paper-23

———, "Objections to a Standing Army (Part I)," Anti-Federalist Paper No. 24, The Federalist Papers Project, 1788b. As of February 8, 2016:
http://www.thefederalistpapers.org/antifederalist-paper-24

———, "Objections to a Standing Army (Part II)," Anti-Federalist Paper No. 24, The Federalist Papers Project, 1788c. As of February 8, 2016:
http://www.thefederalistpapers.org/antifederalist-paper-25

Bull, Hedley, *The Anarchical Society: A Study of Order in World Politics*, New York: Columbia University Press, 1995.

Brunstetter, Daniel, and Megan Braun, "The Implications of Drones on the Just War Tradition," *Ethics and International Affairs*, Vol. 25, No. 3, 2011, pp. 337–358.

Chamayou, Gregoire, *A Theory of the Drone*, New York: The New Press, 2013.

Chambers, John Whiteclay II, *To Raise an Army: The Draft Comes to Modern America*, New York: The Free Press, 1987.

Clausewitz, Carl von, *On War*, ed. and trans. Michael Howard and Peter Paret, Princeton, N.J.: Princeton University Press, 1984.

Cockburn, Andrew, *Kill Chain: Drones and the Rise of High-Tech Assassins*, New York: Henry Holt and Co., 2015.

Coffman, Edward M., *The Old Army: A Portrait of the American Army in Peacetime, 1784–1898*, Oxford, UK: Oxford University Press, 1986.

———, *The Regulars: The American Army, 1898–1941*, Cambridge, Mass.: Harvard University Press, 2007.

Cole, Ronald H., Walter S. Poole, James F. Schnabel, Robert J. Watson, and Willard J. Webb, *The History of the Unified Command Plan, 1946–1993*, Washington, D.C.: Joint History Office, 1995. As of December 3, 2015: http://dtic.mil/doctrine/doctrine/history/ucp.pdf

Craig, Gordon, *The Politics of the Prussian Army, 1640–1945*, Oxford, UK: Oxford University Press, 1978.

Doughboy Center, "In Their Own Words: The Story of AEF by Its Members, Allies and Opponents in Seven Parts," *The Story of the American Expeditionary Forces*, Worldwar1.com, undated. As of December 3, 2015: http://www.worldwar1.com/dbc/ow_2.htm

Editorial Board, "Preparing for Warfare in Cyberspace," *New York Times*, April 28, 2015.

Entous, Adam, "Obama Kept Looser Rules for Drones in Pakistan," *Wall Street Journal*, April 27, 2015.

FitzGerald, Ben, and Parker Wright, *Digital Theaters: Decentralizing Cyber Command and Control*, Washington, D.C.: Center for New American Security, April 2014. As of February 11, 2016: http://www.cnas.org/sites/default/files/publications-pdf/CNAS_DigitalTheaters_FitzGeraldWright.pdf

Franks, Tommy, *American Soldier*, New York: Regan Books, 2004.

Fukuyama, Francis, *The Origins of Political Order: From Prehuman Times to the French Revolution*, New York: Farrar, Straus and Giroux, 2011.

Fung, Brian, "Elon Musk and Stephen Hawking Think We Should Ban Killer Robots," *Washington Post*, July 28, 2015. As of January 7, 2016: https://www.washingtonpost.com/news/the-switch/wp/2015/07/28/elon-musk-and-stephen-hawking-think-we-should-ban-killer-robots/

Gates, Robert, *Duty: Memoirs of a Secretary at War*, New York: Alfred A. Knopf, 2014.

Graham, Bradley, *By His Own Rules: The Ambitions, Successes, and Ultimate Failures of Donald Rumsfeld*, New York: Public Affairs, 2009.

Greenfield, Kent, ed., *Command Decisions*, Center of Military History, United States Army, Washington D.C.: Government Printing Office, 1958.

Grimsley, Mark, "Master Narrative of the American Military Experience: The American Military History Narrative: Three Textbooks on the American Military Experience," *The Journal of Military History*, July 2015, pp. 798–802.

Grossman, Dave, *On Killing: The Psychological Cost of Learning to Kill in War and Society*, New York: Back Bay Books, 2009.

Hackett, John, *The Profession of Arms*, New York: Macmillan Publishing Co., 1983.

Hamilton, Alexander, "The Necessity of a Government as Energetic as the One Proposed to the Preservation of the Union," Federalist Paper No. 23, Library of Congress, 1787. As of February 8, 2016: http://thomas.loc.gov/home/histdox/fed_23.html

———, "The Powers Necessary to the Common Defense Further Considered," Federalist Paper No. 24, Library of Congress, 1787. As of February 8, 2016: http://thomas.loc.gov/home/histdox/fed_24.html

———, "The Same Subject Continued: The Powers Necessary to the Common Defense Further Considered," Federalist Paper No. 25, Library of Congress, 1787. As of February 8, 2016: http://thomas.loc.gov/home/histdox/fed_25.html

Hamre, John, "The Electronic Pearl Harbor," *Politico*, December 9, 2015. As of January 18, 2016: http://www.politico.com/agenda/story/2015/12/pearl-harbor-cyber-security-war-000335

Hennigan, W. J., "Air Force Hires Civilian Drone Pilots for Combat Patrols; Critics Question Legality," *Los Angeles Times*, November 27, 2015.

Howard, Michael, *War in European History*, Oxford, UK: Oxford University Press, 1984.

———, *The Invention of Peace: Reflections on War and International Order*, New Haven, Conn.: Yale University Press, 2000.

Huntington, Samuel, *The Soldier and the State: The Theory and Politics of Civil-Military Relations*, Cambridge, Mass.: Harvard University Press, 1985.

———, *Political Order in Changing Societies*, New Haven, Conn.: Yale University Press, 2006.

Johnson, David E., *Modern U.S. Civil-Military Relations: Wielding the Terrible Swift Sword*, Washington, D.C.: National Defense University Press, 1997.

Kaldor, Mary, *New and Old Wars: Organized Violence in a Global Era*, 3rd ed., Cambridge, UK: Polity Press, 2012.

Karsten, Peter, *Naval Aristocracy: The Golden Age of Annapolis and the Emergence of Modern American Navalism*, Annapolis, Md.: Naval Institute Press, 2008.

Kim, Sung Ho, "Max Weber," revised July 31, 2012, in *Stanford Encyclopedia of Philosophy*, Stanford, Calif.: Metaphysics Research Lab, Center for the Study of Language and Information, Stanford University, 2014. As of December 3, 2015: http://plato.stanford.edu/entries/weber/

Kitfield, James, "The Great Draft Dodge," *National Journal*, December 13, 2014. As of December 3, 2015:
http://www.nationaljournal.com/magazine/the-great-draft-dodge-20141212

Kohn, Richard H., *Eagle and Sword: The Federalists and the Creation of the Military Establishment in America, 1783–1802*, New York: The Free Press, 1975.

Kozak, Warren, *LeMay: The Life and Wars of General Curtis LeMay*, Washington, D.C.: Regnery Publishing, 2009.

Kuhn, Thomas S., *The Structure of Scientific Revolutions*, Chicago: University of Chicago Press, 2012.

Lasswell, Harold, *Essays on the Garrison State*, New Brunswick, N.J.: Transaction Press, 1997.

Lee, Wayne E., "Mind and Matter—Cultural Analysis in American Military History: A Look at the State of the Field," *The Journal of American History*, Vol. 93, No. 4, March 2007, pp. 1116–1142.

Locher, James R., III, *Victory on the Potomac: The Goldwater-Nichols Act Unifies the Pentagon*, College Station: Texas A&M University Press, 2004.

Mahan, A. T., *The Influence of Sea Power upon History, 1660–1783*, New York: Dover Publications, 1987.

Marshall, George C., *Memoirs of My Service, 1917–1918*, Boston: Houghton Mifflin, 1976.

McChrystal, Stan, "Beyond the Draft: Rethinking National Service," *Defense One*, November 29, 2015. As of November 29, 2015:
http://www.defenseone.com/ideas/2015/11/
beyond-draft-rethinking-national-service/124022/

McClellan, Edwin, *United States Marine Corps in the World War*, Washington, D.C.: Headquarters, U.S. Marine Corps, 1920.

McPherson, James, *Battle Cry Freedom: The Civil War Era*, Oxford, UK: Oxford University Press, 1988.

National Defense Research Institute, *Sexual Orientation and U.S. Military Personnel Policy: An Update of RAND's 1993 Study*, Santa Monica, Calif.: RAND Corporation, MG-1056-OSD, 2010. As of February 8, 2016:
http://www.rand.org/pubs/monographs/MG1056.html

National World War II Museum, "By the Numbers: The U.S. Military," web page, undated-a. As of December 3, 2015:
http://www.nationalww2museum.org/learn/education/for-students/ww2-history/ww2-by-the-numbers/us-military.html

———, "By the Numbers: World-Wide Deaths," web page, undated-b. As of December 3, 2015:
http://www.nationalww2museum.org/learn/education/for-students/ww2-history/ww2-by-the-numbers/world-wide-deaths.html

Naval History and Heritage Command, "U.S. Navy Personnel in World War II: Service and Casualty Statistics," web page, April 28, 2015. As of December 3, 2015:
http://www.history.navy.mil/research/library/online-reading-room/title-list-alphabetically/u/us-navy-personnel-in-world-war-ii-service-and-casualty-statistics.html

Panetta, Leon, *Worthy Fights: A Memoir of Leadership in War and Peace*, New York: Penguin Press, 2014.

Parker, Geoffrey, *The Military Revolution: Military Innovation and the Rise of the West, 1500–1800*, Cambridge, UK: Cambridge University Press, 2012.

Pellerin, Cheryl, "Rogers: Data Manipulation, Non-State Actor Intrusions Are Coming Cyber Threats," U.S. Department of Defense, November 19, 2015. As of February January 18, 2016:
http://www.defense.gov/News-Article-View/Article/630495/rogers-data-manipulation-non-state-actor-intrusions-are-coming-cyber-threats

Phillips, Michael E., "U.S. Commandos Span the Globe," *Wall Street Journal*, April 25–26, 2015.

Porter, Bruce D., *War and the Rise of the State: The Military Foundations of Modern Politics*, New York: The Free Press, 1994.

Priest, Dana, and William Arkin, "Top Secret America," *Washington Post*, July 19–20, 2010.

Public Law 80-253, National Security Act of 1947, July 26, 1947, as amended through Public Law 110-53, August 3, 2007.

Public Law 99-443, Goldwater-Nichols Depart of Defense Reorganization Act of 1986, October 1, 1986.

Rearden, Steven L., *The Formative Years*, Washington, D.C.: U.S. Government Printing Office, 1984.

———, *Council of War*, Washington, D.C.: National Defense University Press, 2012.

Rostker, Bernard D., *I Want You! The Evolution of the All-Volunteer Force*, Santa Monica, Calif.: RAND Corporation, MG-265-RC, 2006. As of April 10, 2015: http://www.rand.org/pubs/monographs/MG265.html

Sager, John, "Universal Military Training and the Struggle to Define American Identity in the Cold War," *Federal History*, No. 5, January 2013, pp. 57–74.

Schwartz, Moshe, and Joyprada Swain, *Department of Defense Contractors in Iraq and Afghanistan: Background and Analysis*, Washington, D.C.: Congressional Research Service, May 13, 2011, p. 2. As of February 9, 2016: https://www.fas.org/sgp/crs/natsec/R40764.pdf

Secretary of Defense, "Establishment of a Subordinate Unified Cyber Command Under U.S. Strategic Command for Military Cyberspace Operations," memo, Washington, D.C., June 23, 2009. As of February 11, 2016: https://fas.org/irp/doddir/dod/secdef-cyber.pdf

Segal, David R., *Recruiting for Uncle Sam: Citizenship and Military Manpower Policy*, Lawrence: University of Kansas Press, 1989.

Sheehan, James J., *Where Have All the Soldiers Gone? The Transformation of Modern Europe*, New York: Mariner Books, 2009.

Singer, Peter W., "Can't Win with 'Em, Can't Go to War Without 'Em: Private Military Contractors and Counterinsurgency," Brookings Foreign Policy Paper Series, Washington, D.C.: Brookings Institution, September 2007. As of December 3, 2015: http://www.brookings.edu/research/papers/2007/09/27militarycontractors

———, "Counterproductive," *Armed Forces Journal*, November 1, 2007. As of December 3, 2015: http://www.armedforcesjournal.com/counterproductive/

Smith, E. B., "The New Condottieri and US Policy: The Privatization of Conflict and Its Implications," *Parameters*, Winter 2002–2003, pp. 104–119.

Stecha2, "Kill or Be Killed," video posted to YouTube, August 22, 2008. As of February 9, 2016: https://www.youtube.com/watch?v=Tc4h0qcAIpE

Stewart, Richard W., ed., *The United States Army in a Global Era, 1917–2008*, Vol. 2, *American Military History*, Washington, D.C.: Center of Military History, 2010.

Strawser, Bradley, *Killing by Remote Control: The Ethics of an Unmanned Military*, Oxford, UK: Oxford University Press, 2013.

Stuart, Douglas T., *Creating the National Security Act: A History of the Law That Transformed America*, Princeton, N.J.: Princeton University Press, 2008.

TheUSAHEC, "World War II Basic Training," video posted to YouTube, September 10, 2009. As of February 9, 2016:
https://www.youtube.com/watch?v=CVUIdxQVHQI

Toll, Ian W., *Six Frigates: The Epic History of the Founding of the U.S. Navy*, New York: W.W. Norton and Co., 2006.

———, *Pacific Crucible: War at Sea in the Pacific, 1941–1942*, New York: W.W. Norton and Co., 2012.

United States Code, Title 10, Armed Forces, undated.

United States Code, Title 50, War and National Defense, undated.

United States Special Operations Command, *History of United States Special Operations Command*, 6th ed., MacDill Air Force Base, Fla., March 31, 2008.

U.S. Congress, 37th Cong., 3rd Sess. Chapters 74 and 75, Enrollment Act, Washington, D.C., March 3, 1863.

———, conference committee report for Selective Service Act of 1948, Washington, D.C., June 19, 1948. As of December 3, 2015:
http://www.loc.gov/rr/frd/Military_Law/pdf/act-1948.pdf

U.S. Department of Defense, *The DoD Cyber Strategy*, Washington, D.C., April 2015. As of February 8, 2016:
http://www.defense.gov/Portals/1/features/2015/0415_cyber-strategy/Final_2015_DoD_CYBER_STRATEGY_for_web.pdf

Vandiver, Frank E., "Commander-in-Chief—Commander Relationships: Wilson and Pershing," *The Rice University Studies*, Vol. 57, No. 1, 1971, pp. 69–76.

Wall, Andru E., "Demystifying the Title 10-Title 50 Debate: Distinguishing Military Operations, Intelligence Activities and Covert Action," *Harvard National Security Journal*, Vol. 3, 2011, pp. 85–142.

Watson, Cynthia, *Combatant Commands: Origins, Structure, and Engagements*, Westport, Conn.: Praeger Security International, 2011.

White, John P., Antonia H. Chayes, Leon A. Edney, John L. Matthews, Robert J. Murray, Franklin D. Raines, Robert W. RisCassi, Jeffrey H. Smith, Bernard E. Trainor, and Larry D. Welch, *Directions for Defense: Report of the Commission on Roles and Missions of the Armed Forces*, Washington, D.C.: U.S. Government Printing Office, 1995. As of December 3, 2015:
http://edocs.nps.edu/dodpubs/topic/general/ADA295228.pdf

Wilson, James Q., *Bureaucracy: What Government Agencies Do and Why They Do It*, New York: Basic Books, 1989.

Wilson, Woodrow, "Wilson's War Message to Congress," World War I Document Archive, April 2, 1917. As of February 9, 2016:
http://wwi.lib.byu.edu/index.php/Wilson%27s_War_Message_to_Congress

———, "President Wilson's Proclamation Establishing Conscription," Firstworldwar.com, May 28, 1917. As of December 3, 2015:
http://www.firstworldwar.com/source/usconscription_wilson.htm

Zegart, Amy, *Flawed by Design: The Evolution of the CIA, JCS, and NSC*, Stanford, Calif.: Stanford University Press, 1999.